W9-AEO-857
A S Q C
LIBRARY

DESIGNED TO WORK
Production Systems and People

ROBERT T. LUND
Boston University

ALBERT B. BISHOP
The Ohio State University

ANNE E. NEWMAN
Cummins Engine Company

HAROLD SALZMAN
University of Louisville

PTR PRENTICE HALL, Englewood Cliffs, New Jersey 07632

Library of Congress Cataloging-in-Publication Data

Designed to work : production systems and people / Robert T. Lund . . .
[et al.].
p. cm.
Includes bibliographical references and index.
ISBN 0-13-203944-3
1. Production planning. 2. Man-machine systems—Planning.
3. Work design–Case studies. I. Lund, Robert T.
TS176.D47 1993 92-41381
658.5—dc20 CIP

Acquisitions editor: Michael Hays
Editorial/production supervision and interior design: Maria McColligan
Copyeditor: Kathy Dix
Prepress and manufacturing buyer: Mary McCartney
Cover design: Karen Marsilio

Printed in the United States of America
10 9 8 7 6 5 4 3 2 1

ISBN 0-13-203944-3

Prentice-Hall International (UK) Limited, *London*
Prentice-Hall of Australia Pty. Limited, *Sydney*
Prentice-Hall Canada Inc., *Toronto*
Prentice-Hall Hispanoamericana, S.A., *Mexico*
Prentice-Hall of India Private Limited, *New Delhi*
Prentice-Hall of Japan, Inc., *Tokyo*
Simon & Schuster Asia Pte. Ltd., *Singapore*
Editora Prentice-Hall do Brasil, Ltda., *Rio de Janeiro*

CONTENTS

PREFACE

How do you design advanced manufacturing technology to get outstanding performance? Experience has shown it takes more than conventional engineering to create "sophisticated," "integrated," "flexible," "robust," "productive," "reliable," manufacturing systems that really live up to expectations. The *people* who work with the systems determine the performance of machines and their software. Characteristics of the machines, in turn, determine the nature of work for the people associated with the machines. Because this is so, a kind of synergy has to exist between machines and human beings if they are to work well together. This need for compatibility must be recognized when design and/or selection of new technology begins.

This book embodies the collective insights of hundreds of individuals from many of the largest American manufacturing companies. We report the findings of a three-year study into the design and selection of advanced process technology in manufacturing. We examine the extent to which machine–human compatibility is considered early in the acquisition of new production systems. We identify design principles, which, if applied, will speed implementation and increase performance of production systems.

Our objective is to give manufacturing managers a clear set of policies to guide design and selection of new manufacturing technology so they can achieve the full benefits of new systems investments. We show how companies have gained significant advantages by applying these policies, and we show how things can go wrong when machine–human compatibility principles are ignored. Although the subject is technical, we have tried to express ideas in terms that will be readily grasped by anyone in industry. It is as important that production workers understand these concepts as it is that managers understand them.

The lessons in this book are also for engineers and the engineering institutions that train engineers. Our intent is to broaden designers' perspectives. We point out the human factors that influence the success of a manufacturing system. We provide easily understood design principles that have been demonstrated to work. We also add references that allow the interested student or practitioner to pursue a subject in greater depth. Thus, the book can be used as a supplementary text in an engineering design course.

This work was undertaken by two research teams. One team, headed by Albert B. Bishop, Professor of Industrial and Systems Engineering, was located at The Ohio State University. Its members were: Anne E. Newman, Director, Operations Analysis, Cummins Engine Company; Richard Klimoski, Professor of Psychology; John B. Neuhardt, Professor of Industrial and Systems Engineering; Cheryl M. Burgess, Technical Typist; Brian E. Campbell, Graduate Research Associate; Kenneth Fanta, Graduate Research Associate; and Thomas E. Miller, Graduate Research Associate.

The second team was directed by Robert T. Lund, Research Professor of Manufacturing Engineering. It was based at Boston University. This team included: Harold Salzman, Associate Research Scientist and Director, Technology, Work, and Organizations Program, University of Louisville; Elias Zahavi, Professor of Mechanical Engineering, Ben Gurion University of the Negev (on sabbatical to B.U.); Elizabeth Ehrenberg, Graduate Research Assistant; Abhijeet Ghatak, Graduate Research Assistant; Virender Sandhu, Graduate Research Assistant; and Tamara Upham, Graduate Research Assistant.

At the time of the study, Anne Newman was a Graduate Research Associate studying for her doctorate at The Ohio State University, and Harold Salzman was an Adjunct Assistant Professor at Boston University.

The book is the fulfillment of career-long objectives of the two team leaders, who have devoted much of their professional lives to preparing engineers for positions in industry. It was a major undertaking for their two associates. The book could not have been written, however, if it were not for the assistance and good will of many others.

The sponsor of the research project was the National Science Foundation's Manufacturing Processes and Equipment Program. We are grateful for the support provided by NSF and wish especially to thank Dr. Marvin deVries, now at the University of Wisconsin, for his consistent encouragement.

Our survey results were made possible by the high degree of cooperation we received from executives of leading manufacturing companies in this country. We are grateful for the hours given by these executives in answering our many questions. We agreed to keep the names of the respondents and their companies confidential, but each of the participants should feel that he or she has contributed significantly toward a better understanding of human factors policies and practices in machine design.

We are especially appreciative of the cooperation from five manufacturing organizations: Camera Division of Polaroid Corporation, Sikorsky Aircraft Division of United Technologies Corporation, Faircrest Steel Plant of Timken Company, Westinghouse Electrical Systems Division of Westinghouse Electric Corporation (now a part of Sundstrand Corporation), and Clyde Division of Whirlpool Corporation. Each of these firms allowed us to conduct case studies of the design (or selection) and installation of major new manufacturing systems. The studies, presented in Chapters 5 through 9, provide a wealth of information on the process of acquiring new process technology.

Although we acknowledge the invaluable assistance we have received from industrial participants and members of the research teams, the findings reported herein are our responsibility. They do not necessarily represent the views of the National Science Foundation, Boston University, or The Ohio State University.

R.T.L.
A.B.B.
A.E.N.
H.S.

Note: "Polaroid" and "Spectra" are trademarks of the Polaroid Corporation. "Timken" is the trademark of the Timken Company.

1

INTRODUCTION

We were talking to members of the team that had successfully completed design and installation of a major computer-based manufacturing system. Productivity, quality, costs, yields, and capacity had all been enormously improved. From the time the design concept for the machine system was first established, experienced operators had been involved in assessing the design and making suggestions. Many of their ideas had been adopted.

"You'd be surprised at how important communications are in a project like this," one of the design leaders said. "We thought we had been completely up front with the operators, but we still had problems.

"After the equipment had been installed and everything seemed to be working properly, we began to have sudden stoppages of the system. Everything would go dead. We couldn't find an explanation. Nothing seemed to be wrong, yet the machine kept stopping.

"This went on for two weeks. It was driving us crazy. Then we learned, through the grapevine, what was happening. One of the operators had realized how productive the new system would be, and he figured that there would have to be layoffs as a result. He was quietly pushing in the stop button for the whole system and then immediately pulling it out again. The system would come to a halt, and there was no clue as to what had gone wrong.

"We had found the source of our problem, but now we were uncertain what to do about it. Fortunately, while we were still debating the issue, the operator became convinced that we really were not going to lay anyone off. The stoppages ceased, and the system has run well ever since.

"The story doesn't end there. We were so impressed with this person's understanding of the system that we encouraged him to get more education. Now he's one of our manufacturing engineers, trouble-shooting *real* machine problems."

Battles for economic supremacy among the most advanced nations of the world are being waged on the manufacturing front. Countries whose products are unable to compete on a global basis are experiencing serious trade imbalances and slowing economic growth. Two of the major battle lines in this international competition are closely related—product innovation and advanced process technology. From the American point of view, product innovation, the creation of new products and the upgrading of existing products, has long been a major strength. New products are introduced at a prodigious rate. Product life cycles are becoming shorter, reflecting an acceleration of that rate.

The adoption of advanced process technology in the United States, on the other hand, tends to lag behind product innovation. The consequence has been that new products are introduced by American manufacturers but are frequently appropriated by offshore producers, who proceed to take over the world market. The capture of the American-designed videocassette recorder by Japanese electronics manufacturers is a typical example. The Japanese advantage in that instance was not because of lower labor rates; it was due to superior manufacturing capability, resulting in far higher productivity rates and consistency of quality.

Regardless of its innovative capabilities in product development, any country that cannot sustain competence in manufacturing faces serious economic consequences. Unless American manufacturers can rapidly improve their production capabilities, much of the innovative effort that goes into product development will be wasted. This is why the adoption of advanced manufacturing technology has become a major issue in the United States.

The desire of American firms to use the newer manufacturing technologies appears to be strong. According to a report from the Department of Commerce (Bureau of the Census, 1989), of the nearly 40,000 establishments in the United States engaged in manufacturing durable goods, 68 percent were already using at least one of 17 advanced manufacturing technologies by late 1988. Almost 60 percent of these firms expected to add at least one more technology in the next five years. The usage rates for many of these recent process developments was expected to double from 1989 to 1993.

The types of new process technologies now available to industry are highly dependent on electronic controls—computers or programmable controllers. The process may involve cutting, forming, joining, finishing, assembling, mixing, or reacting. It may combine several of these elements with a handling system that moves parts, tools, and fixtures to and from locations in the system. Whether the new technology involves cutting tools, lasers, electron beams, plasmas, robots, extruders, presses, welders, sprayers, or any of a multitude of other process alternatives, some form of electronic programmable logic is likely to be part of the system. It is the application of computers or computer-like controls that has made such a profound difference in the nature of manufacturing technology.

The means and the know-how to create superior manufacturing systems are now fully available and have been demonstrated in many applications. Robust computer systems; materials transfer and handling devices; automated, precise machines; and a diversity of sensing techniques are accessible to industrial firms in all product markets. In the 1970s it was reasonable to say that advanced technology had not yet reached the point where it could perform reliably and economically. In the 1990s this excuse is no longer valid. If companies are lagging behind in the use of new process technology, the reasons must be found outside the technology itself.

Not only has the rate of acceptance of these new technologies been slow, but their effectiveness, once they have been installed, has often fallen short of expectations. Companies that have been disappointed in the outcomes of these investments are beginning to realize that technological changes cannot be treated simply as technical problems. Putting advanced systems to work in industry involves new approaches to the design, selection, and implementation of such systems.

To employ advanced manufacturing technology, firms must first design or select an appropriate process or system; then they must make it work. Even after machines are running, the implementation effort is far from complete. Consistent, reliable production must be attained. Further, processes must be continually upgraded and improved to reach the maximum benefits from these increasingly more complex and expensive systems.

ENGINEERS DESIGN JOBS, NOT JUST MACHINES

When engineers set out to design manufacturing systems, they tend to be driven by three major factors:

- Time constraints, such as delivery schedules, lead times, product introduction dates, and product throughput times.
- Technical feasibility considerations that guide process selection decisions (i.e., what technology will give the desired results?).
- Financial and economic considerations such as capital investment, installation costs, unit production costs, overhead costs, profitability, and return on investment.

In addition to these general considerations, system designers work to a set of criteria on performance measures by which the process will be evaluated. These criteria include machine capability, speed, accuracy, capacity, reliability, flexibility, delivery, and cost. Largely because of the legal imperatives of the Occupational Safety and Health Act (OSHA), designers have also been required to add worker health and safety to their list of criteria, and some also include consideration of the ergonomic limitations of humans.

Ultimately, when new manufacturing systems reach the factory floor, people—operators, mechanics, technicians—must work on and with them. Except for matters of safety, health, and ergonomics, however, engineers have not usually been concerned with designing manufacturing systems to take into account the full range of human capabilities and limitations. This aspect has not been considered a part of their responsibilities, and they have largely been content to deliver a completed manufacturing system to the production floor. They have counted on manufacturing supervisors and trainers to introduce workers to the new machines and to help them accommodate to the changes.

The fact is, however, that the design of a machine largely determines the nature of the job of the person who will work with it. Once the configuration, the mode of control, the method of loading and unloading product, the operational sequences, the tools, and the adjusting mechanisms have all been established and frozen into hardware and software, the nature of the operator's job has been determined. Engineers, therefore, do not merely design machines. They design jobs, whether they are aware of this or not.

MOTIVATION FOR THIS STUDY

If the adoption of new process technology by American firms has been slow and disappointing, and if the designers of that technology are only minimally involved in the human aspects of machine operation, could there be a cause-and-effect relationship between the two circumstances? When we undertook the research that is the basis of this book, we believed there was.

We believed that if there were a relationship, we might be able to discover principles for the design and selection of advanced manufacturing systems that would guide systems engineers to more productive and more readily accepted technology. The hypotheses we set out to prove were these:

1. The criteria and measures of effectiveness (policies) currently used to guide manufacturing systems design and selection generally do not include human factors other than ergonomics.
2. Principles of machine–human compatibility can be derived from analyses of case studies of the design and implementation of actual operating systems.
3. The principles thus derived can guide future system design and selection.
4. If production systems are designed according to these principles of machine–human compatibility, benefits will accrue both to the firm and to the worker.

FINDINGS, IN BRIEF

It is our conviction, now reinforced by the research reported in this book, that there is indeed a relationship between system effectiveness and the degree of concern for human factors manifested in the design and selection process. By taking into account the full range of human capabilities, systems designers can create manufacturing technology superior in performance and in acceptance to that developed conventionally.

In subsequent chapters we report a substantial amount of detail from the three-year project. The major findings, however, are these:

- Despite their significance, there are relatively few human factors policy areas that apply to the design and selection of manufacturing systems. We identified only 12. Of these, five pertain to physiological needs: health, safety, comfort, stress, and ergonomics. The remaining seven policy areas relate to the cognitive capabilities of the individual:

  1. Involvement in the design or selection process
  2. Control of the process
  3. Feedback of production information
  4. Modification of the process
  5. Skills utilization
  6. Maintenance responsibility
  7. Participation in postinstallation design change

- With the exception of the physiological aspects of health, safety, and ergonomic design dictated by federal law, a majority of the leading industrial firms we surveyed had few policies relative to human factors that applied to the design and selection stages of equipment acquisition.
- Relatively few firms, even among the largest in the country, design more than 25 percent of their manufacturing systems. A very small minority build more than 25 percent of the equipment they use.
- Elimination of labor or reduction of labor's influence is a major motivator in the acquisition of new technology for firms in the United States.
- Difficulties encountered in implementing new manufacturing technology tended to fall into two categories: problems during installation (time, cost, performance), and lower-than-expected system performance after installation. Our survey results and case studies indicate that both types of problems can be reduced if the ultimate users of the new technology are directly and consistently involved in the design and selection process. Equipment performance is further enhanced if systems are designed to facilitate operator control, modification, and maintenance of the process, and if company policies encourage employee participation in programs to improve process performance.
- The work environment influences the motivation of people to learn and use new skills. Design efforts to "deskill" jobs or to lock the operator out of the process tend to create problems related to machine downtime and product quality, because the operator is neither trained nor permitted to intervene when problems arise.
- Effective use of employee skills depends on timely, accurate feedback of oper-

ating results. Computer-aided information systems that give operators first-hand, real-time data on product and process conditions are important tools that enable operators to control the process.

- Except in instances where union agreements interfere, responsibility for first-level maintenance increasingly needs to be a part of the operator's responsibility. Operators can keep expensive systems running productively if they are given the appropriate training and tools, and if their machines are designed so they can intervene easily and safely.
- Full use of a person's skills in operating or maintaining a manufacturing system involves provision for training and development, job rotation, opportunity for advancement, and recognition for contributions made toward process improvement.
- Except for a trend toward greater inclusion of topics related to the physiological factors, writers of machine design textbooks over the past 50 years have generally excluded consideration of the human beings who will be working with the machines.

IMPLICATIONS OF THE FINDINGS

There is a significant blind spot in industrial policy and practice in this country. If all the resources available to a firm are to be fully utilized, people and technology must be teamed, not separated. Industrial leadership must broaden its approach toward technological change. Companies cannot continue to focus exclusively on technical aspects of manufacturing productivity. There is a need for more universal recognition that manufacturing processes involve both people and machines, even in situations where automation has been carried to the furthest extreme.

Operating management must convey this recognition to the engineers who design and select new process technology for a firm. It must set policy guidelines and insist on practices that follow principles such as are enumerated at the conclusion of this report.

A change in culture is involved. Engineers are not trained to think in terms of the cognitive skills that humans bring to a job. Such training as they may have relative to humans gives engineers a view of people as machines endowed with certain mechanical and sensory capabilities and limitations. Encouraged by management philosophies that regard labor as a cost to be eliminated, or as a threat to management's independence, engineers have sought to automate whatever can be automated and to subdivide the balance of work into readily learned tasks requiring little skill. The possibility that workers might contribute significantly to the production process has not been part of the engineer's training or of his or her industrial experience. We found this to be the case in many of the leading firms we surveyed.

The change to designing machines as if people really matter must come with full management support. In situations where management and labor relations have been hostile, a step toward involvement of workers in process design and selection may be met with skepticism or suspicion. Participants will be trying to uncover management's hidden motives.

If the changes are made honestly and openly, with effective training and communication, our evidence indicates that new technology will not only be employed more readily, but it will often perform at levels beyond the expectations of its designers. Further, continued improvement in the performance of new manufacturing systems can be anticipated when workers are made knowledgeable and are motivated by an early association with the new technology.

The institutions that are training our engineers must likewise undergo a cultural change. Engineering curricula of our universities have little to do with human capabilities. Such matters are certainly not in the core courses required of future systems designers.

If one key to the success of advanced manufacturing technologies is to incorporate human capabilities into system design, then the understanding of these capabilities must be incorporated into the intellectual core of our engineering schools. This task may well be more difficult to achieve than making the changes in industry. Engineering faculty are generally ill equipped to deal with the subject. Even the younger engineering faculty, with their keen appreciation of computers and advanced design techniques, are no better equipped to deal with human factors than are faculty who were trained back in the age of mechanization.

Engineering schools may also insist that their curricula are already too crowded; that they cannot teach all that is necessary within the time allotted for a degree program. The possibility of having to add more subjects poses a difficulty. The only encouragement we have to offer is the fact that the principles we propose are simple. They are consonant with commonly accepted views of human nature and human capability. The introduction of these ideas will not require new courses in psychology or industrial sociology either for faculty or students. The precepts can, and should, be incorporated into regular machine design or manufacturing systems design courses. In this way the principles will have equal standing with other design principles and will not be regarded as secondary in importance.

There are engineers in industry who have learned these principles and are successfully applying them. It will be possible, therefore, for those who write machine design textbooks to find useful examples of how the principles are employed. Once the concepts have been embodied in standard texts, it will be considerably easier (and more acceptable) for engineering faculty to devote classroom time to the human side of the machine–human compatibility issue.

WHO SHOULD BE CONCERNED?

Industrial management, seeking ways to gain global competitive advantage, should be concerned about the findings of this study, and encouraged by the examples of success found in some of our case studies. Changes in design and selection policies can come about only through executive action. In the absence of clear directions from this quarter, engineering departments will continue to design and select new technology as they have for decades.

Engineers who are puzzled as to why their designs are less successful than anticipated may find some new answers among the ideas introduced in this report. To some extent, they can initiate changes in their approaches to new systems acquisitions, even if policy guidelines are missing. Changes in attitude are at the root of many of the principles that are recommended. Humility that recognizes the worth of a production employee's idea is one such attitudinal change.

Educators have an important, and as yet unrecognized, stake in this situation. Those who prepare engineers for professional careers and those who write the textbooks are both responsible for ensuring that designers in the next generation are trained in all aspects of their art. As has already been mentioned, however, changes from the status quo in education tend to be glacial.

The people who have the most to gain from the changes advocated here are the industrial workers who must deal with whatever technology the engineers design or buy. These are the people whose lives are directly impacted by the nature of the work they are asked to do and the environment in which they perform this work. A person who is first employed at the age of 20 and retires at 65 spends at least 90 thousand hours on the job—an enormous part of that individual's life.

Finally, American society is concerned about the economic future of a country that seems unable to sustain its domestic manufacturing capability. Further decline of this segment of our economy would have serious consequences relative to the availability of jobs and our power to maintain our standard of living.

To a greater or lesser degree, then, everyone has a stake in the issues that are explored here.

ORGANIZATION OF THIS BOOK

In the subsequent chapters of the book we first describe the results of our national survey of leading manufacturers. The survey sought to determine current policies and practices regarding human factors in the design and selection of manufacturing technology. The findings of that survey include identification of the twelve human factors policy areas relevant to system design. We next report on our examination of process design texts to show the extent to which the experts have considered human factors as part of the necessary training of designers.

We then present five case studies of major advanced system installations in five leading manufacturing firms. These case studies are in sufficient detail to identify the human factors design principles used. The cases are then analyzed for similarities and contrasts in approaches and in outcomes.

In the final chapter we present a set of design principles derived from the study that should become part of the armamentarium of any manufacturing system designer. These 15 simple principles, if followed by designers and supported by company policies, will return substantial rewards for any added efforts that may be required. Suggestions and examples of how these principles may be applied are provided.

REFERENCE

Bureau of the Census, (May, 1989). U.S. Department of Commerce, *Manufacturing Technology 1988.* Washington, DC: U.S. Government Printing Office.

2

SURVEY DESIGN AND RESULTS

RESEARCH DESIGN

Research teams at Boston University and The Ohio State University conducted telephone interviews with senior manufacturing executives in leading manufacturing firms. The target firms were those that ranked, in terms of market share, among the top three in each of the major Standard Industrial Classification categories in the metalworking industries (SIC 34 through 38). Products made by these firms ranged widely, including aircraft, automobiles, instruments, machinery, photographic equipment, household appliances, and many others. Letters soliciting participation were sent to 180 firms in these categories. Sixty-three companies, 35 percent of the sample, agreed to participate.

In some instances, the "companies" that were selected for predominance in a given market were actually divisions of larger diversified companies. In a few cases, this resulted in our interviewing more than one division of a firm. Other complications arose in some cases when it was found (1) that the company or division chosen had been sold or acquired by another firm, or (2) that the person selected by company management to respond was from a manufacturing division different from the one we had selected. In either of these situations, we proceeded with the interview if the manufacturing activities were considered relevant to the study. As a consequence, two firms, one in specialty chemicals and another in primary metals fabrication, were included in the survey, even though their products did not fall within the planned SIC range.

The interviewees were vice-presidents of manufacturing or technology, directors of engineering or operations, and manufacturing managers. In a number of instances, we actually interviewed teams of people from firms by means of conference calls. A typical team might include the vice-president of manufacturing, a manufacturing engineering manager, and a manager of equipment design.

The telephone interviews, guided by a questionnaire, ranged from one hour to over two hours in length, with most interviews taking about one and one-half hours. We first obtained background information on company size, sales, and employment, as well as on market competition and concentration. In addition, we inquired about the company environment with respect to attention paid to tasks versus attention paid to people. For this we used a perceptual measure, asking the interviewee to rate, using a nine-point scale (1 to the lowest, 9 to the highest), the level of company concern for getting tasks completed. Then, on the same scale, we asked the respondent to rate the company's concern for people. We used these measures to assess the relative concern for tasks versus people by computing a ratio of people-to-task ratings.

We next asked about the company's equipment design practices. We inquired about the percentage of their manufacturing equipment that was designed in-house, and the amount that had been built in-house. For companies that had little in-house design we focused on their procurement process and the specifications they set for their suppliers of equipment. We then determined what groups of people in the firm participated in the design or selection process, the role each played, and the organization of their engineering staff (e.g., whether it was a centralized or decentralized activity).

The next part of the survey asked about human factors policies and principles considered during the design or selection of manufacturing equipment. We first asked the respondents to identify any human factors policies the company had, and to state whether each policy so identified was "explicit" or "implicit." To be considered explicit, a policy had to be a written company document; an implicit policy had to be widely understood and practiced, even if unwritten. Policies of either kind had to be recognized and enforced, informally or formally, by leaders or managers of the design groups. We also recorded information and specific practices or projects involving human factors, but we did not regard these as company policies unless the underlying rule met either of the above policy criteria.

After allowing the respondents to volunteer information about their companies' human factors policies and practices, we then asked about specific policies from a list of human factors areas that had been tentatively assembled by the research team. For each area in which the company was claimed to have a policy, we asked the respondents to describe the policy and how it was applied in their equipment design or selection projects. This part of the survey elicited rich detail about the nature of these policies and their applications.

The final part of the interview was a detailed set of questions about two specific design or selection projects in the company with which the interviewee was familiar. For each project, the interviewee was asked to provide a background description of the equipment design project, design procedures used, design policies and principles employed, installation times, outcomes in system performance and reactions of the labor force.This part of the survey provided the most in-depth information about actual design practices, equipment design and operation, and outcomes. Information about these projects was most valuable for evaluating the extent to which human factors policies were employed and the impact on production operations.

In the following sections of this chapter we present our survey findings. These include background data on the companies in the survey, their product markets, types of production processes, and their equipment acquisition practices. We then present data on the outcomes of specific cases of adoption of new manufacturing technology, analyzing these outcomes in terms of system performance and impact on the work force. Finally, we present and analyze the human factors policies identified through the survey as being in effect in one or more firms during equipment design or selection.

COMPANY BACKGROUND

Companies in the survey were from SIC industries 34, 35, 36, 37, and 38 (fabricated metal products, machinery, electrical and electronic machinery, transportation equipment, and instruments). Fifty percent of the companies were the largest manufacturer in their product area, and another 27 percent were the second or third largest in their area. These were large companies, with over 50 percent of the sample having sales of over $800 million (Figure 2–1). The median employment for this sample of firms was 10,000. Twenty-five percent of the firms had over 25,000 employees.

Most of the companies had a substantial or dominant share of their product market. Nearly a quarter of the companies held more than 50 percent of the market

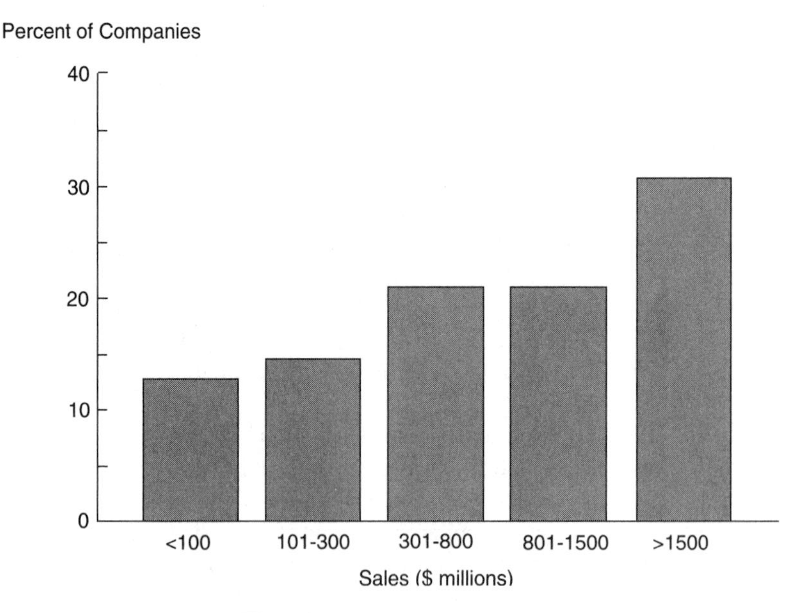

Figure 2–1 Annual company sales

share in their area; a third had 26 to 50 percent of their market. These companies competed in highly concentrated industries, with over half (57 percent) of the companies stating that five or fewer firms controlled 75 percent of their product market. Only 11 percent of the respondents were in markets where sales were dispersed among 10 or more firms. Nevertheless, 94 percent of the respondents reported that there was strong competition in their market area.

Product Attributes

The product markets of these companies were widely distributed among industrial, commercial, consumer, and military customers (see Table 2–1). Products of half of the companies had long product life cycles (over 10 years), and 38 percent had median life cycles of 4 to 10 years, with only 10 percent having short life cycles of 3 years or less.

When asked to list key attributes of their products, that is, the characteristics of their products giving them a competitive edge, 51 percent of the respondents identified product quality. Thirty-five percent of the firms listed product features as competitively important. Product capacity, low cost, customer service, reliability, and high performance were each listed by fewer than a quarter of the companies (see Figure 2–2).

Company Environment

We used a perceptual measure to gauge overall company concern for accomplishing tasks and overall company concern for employees. We asked the interviewees to rate their company on a "task concern" scale and then on a "people concern" scale, both ranging from 1 to 9. Respondents were first asked, "On a scale of 1 to 9, nine being the highest score, how do you rate your company in terms of concern for getting things done?" The second question was, "On the same scale, how do you rate your company in terms of concern for people in the firm?" The mean rating for task concern was 8.1 and the mean rating for people concern was 6.8, giving a mean

TABLE 2–1 Distribution of Sales by Market Sector

	Percent of Companies				
Sales in Sector	Industrial	Commerical	Consumer	Public	Military
0 percent	57%	53%	64%	91%	75%
1 to 30 percent	14%	22%	10%	8%	4%
31 to 70 percent	8%	5%	11%	2%	5%
71 to 100 percent	21%	21%	16%	0%	18%

Example: For 22% of the companies, sales in the commercial sector equaled from 1% to 30% of their total sales.

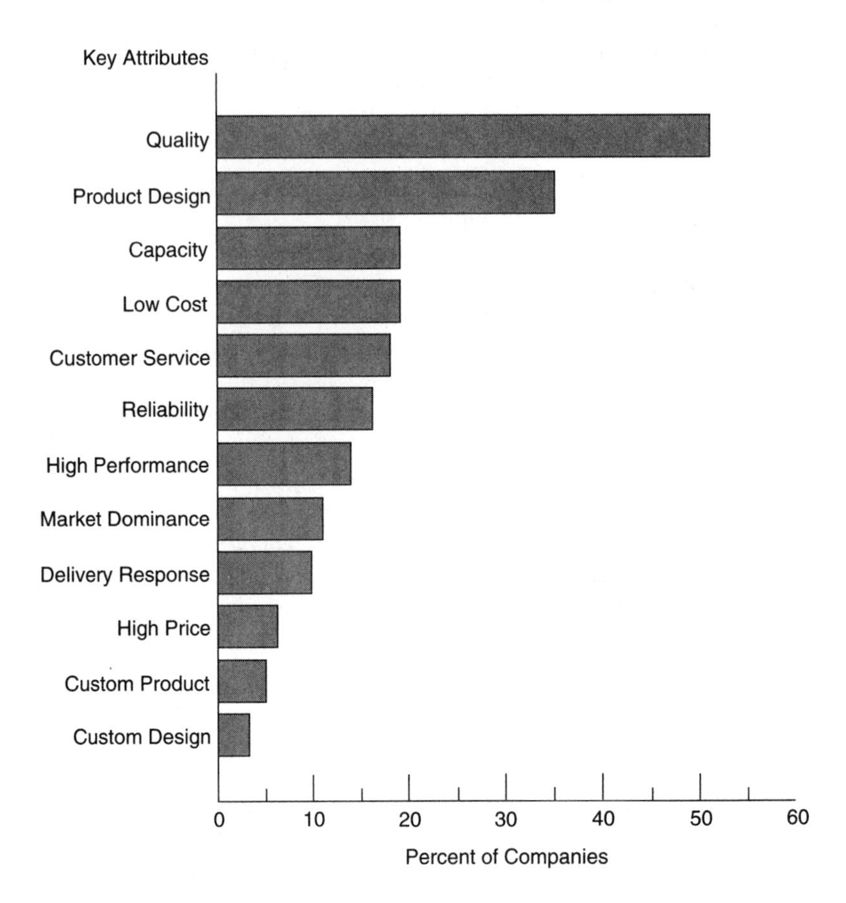

Figure 2–2 Product attributes of companies in sample

"people-to-task ratio" of 84 percent. Seventy-three percent of the companies had a higher task than people concern, 16 percent had equal task and people concern, and only 11 percent had a higher people than task concern (Figures 2–3 and 2–4).

Production Equipment

The firms we interviewed generally were investing large sums in new production equipment. All but 2 of the 59 companies responding to our inquiry on investment had spent over $10 million in new equipment over the previous three years (Figure 2–5). Ten of the firms had invested over $100 million in that time period.

The companies varied widely in the percentage of their production equipment that was computer based, or programmable. About two-thirds reported that up to 50 percent of their equipment was computer based or programmable, 18 percent of the companies had half to three-quarters of this type of equipment, and another 18 percent had more than 75 percent programmable or computer-based equipment.

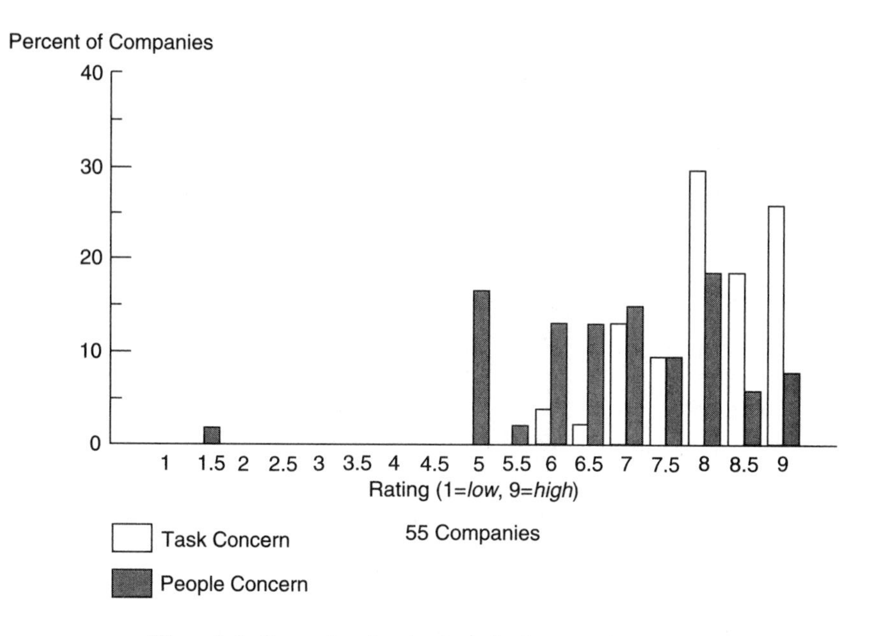

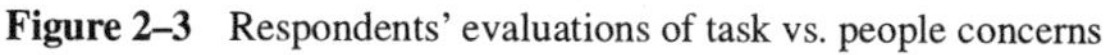
Figure 2–3 Respondents' evaluations of task vs. people concerns

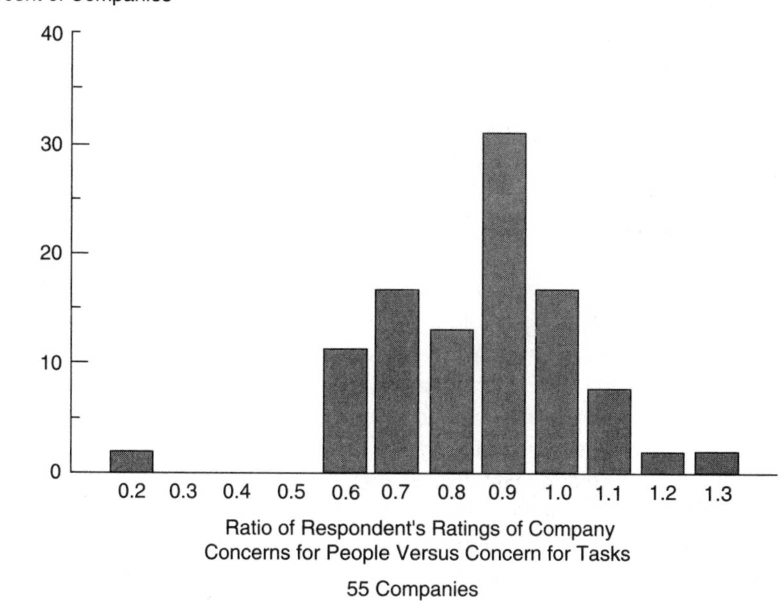

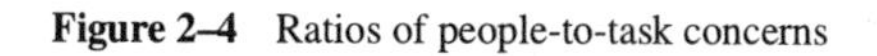
Figure 2–4 Ratios of people-to-task concerns

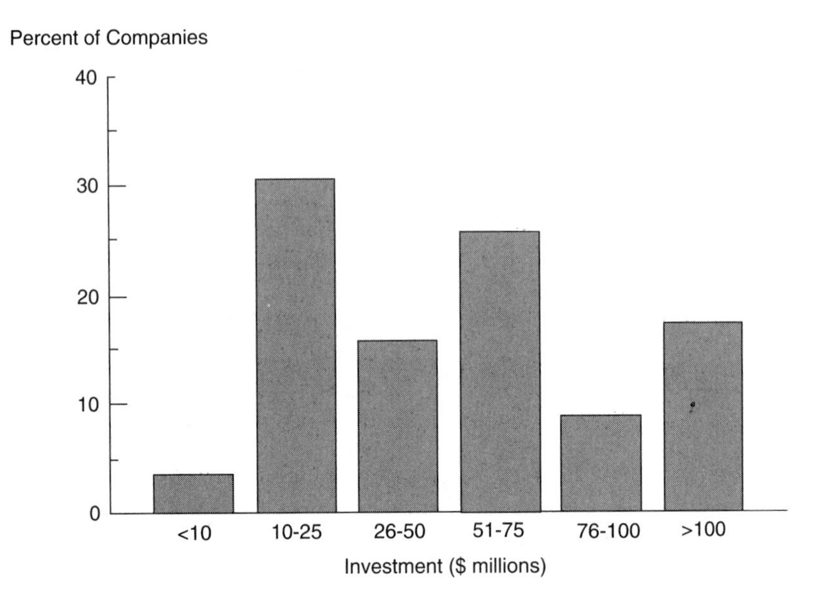

Figure 2–5 Investment in production equipment over previous three years

Most of the companies designed only a small portion of their production equipment in-house (Figure 2–6). About one-quarter of the sample designed none of their equipment. Of the companies that did design some of their equipment, 38 percent designed less than half, and only about 15 percent designed more than 75 percent of their equipment. The figures were even more skewed regarding building their own equipment. Forty-five percent built no equipment in-house, and another 46 percent built less than half their equipment in-house. Only 3 percent (two firms) built 75 percent or more of their equipment.

Experts have argued that firms seeking to enhance their competitiveness should focus on quality improvement at least as much as on cost reduction. Acquisition of new production equipment can be an important aid for achieving improved quality. Our survey confirms the acceptance of this point among the leading manufacturers. Quality leads the list of reasons for adopting new equipment. It was the reason noted by 64 percent of the companies, with productivity, flexibility, and cost savings listed by 41, 33, and 32 percent of the companies, respectively (Table 2–2).

Formal justification for equipment acquisitions, however, required an appropriate Return on Investment (ROI) expectation in a very substantial majority of the firms. ROI was the predominant justification requirement in 79 percent of the companies (Table 2–3). A number of interviewees commented on this divergence between the operating necessity to upgrade equipment to improve product quality and the formal company requirements of ROI justification. These interviewees often felt that ROI justification requirements inhibited the upgrading of equipment necessary to improve their products. Although quality improvement was necessary for the com-

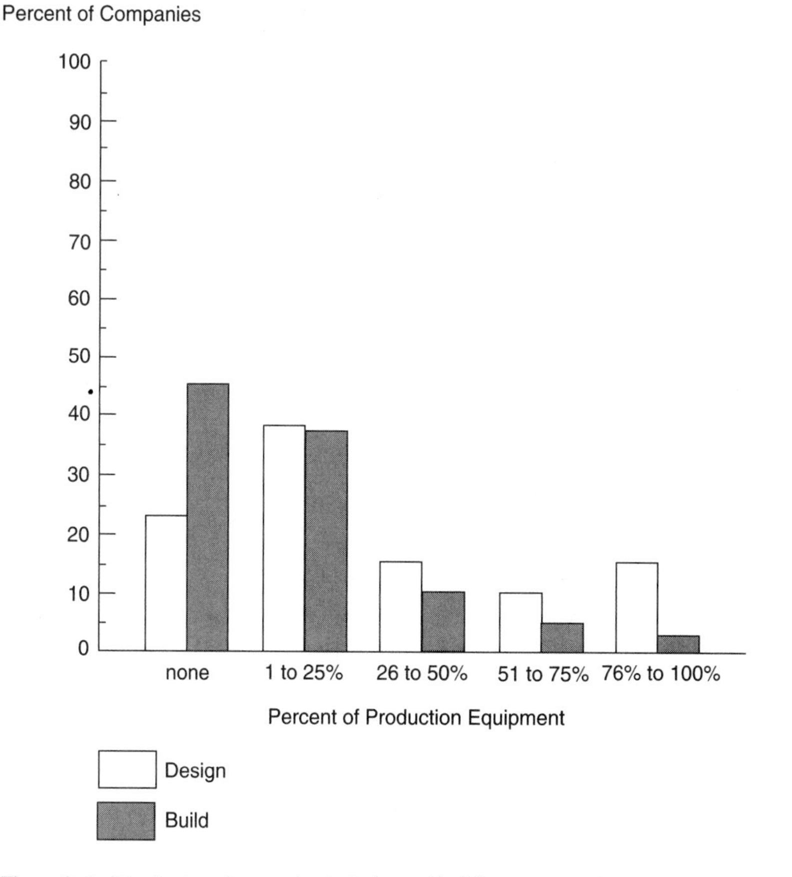

Figure 2–6 Distribution of companies designing and building own manufacturing equipment

pany to remain competitive, it was not always possible to justify equipment purchases in terms of strict ROI criteria.

This ambivalence between need and justification criteria was recognized as a problem by interviewees. At one company, for example, the reasons for adopting new technology were "flexibility, self-diagnostics, statistical process control and analysis capabilities," but, for the purpose of justification, the company required "direct labor reduction". An interviewee from another company commented that the primary justification was "ROI; everything here is investment driven." The representative of a manufacturer of large transportation equipment stated that their primary adoption reason was to "eliminate human intervention and cut labor costs," with ROI as the criterion for justification. In postinvestment evaluations in this firm they ask simply, "Are the people gone?"

Although ROI and labor savings were the predominant approaches to new equipment procurement justification, there were a number of companies that were

TABLE 2–2 Reasons for Adopting New Production Equipment

Reason	Number of Companies	Percentage
Quality	40	63.5
Productivity	26	41.3
Flexibility	21	33.3
Cost savings	20	31.7
Information-tracking capabilities	14	22.2
Low flow time	11	17.5
Accuracy	10	15.9
Reliability	8	12.7
Only reasonable approach	8	12.7
Needed for production requirements	5	7.9
Reduce maintenance	4	6.3
Ease training	3	4.8
Customer required	3	4.8
Fast installation	2	3.2
Ease of maintenance	2	3.2

changing their traditional approaches. The exclusive focus on ROI justification was noted by an interviewee in one large manufacturing company as narrowing their view of the factors that provide production benefits: "Higher quality and less rework reduce costs. Productivity is tied to quality, but we didn't recognize this previously." Another said that in "total Computer Integrated Manufacturing the payback becomes 'softer'—less apparent, less quantifiable. For example, lower inventory, lower buffer stock, and short cycle times are all hard to justify."

TABLE 2–3 Justification Criteria for New Equipment

Reason	Number of Companies	Percentage
ROI	50	79.4
Quality	22	34.9
Needed for production requirements	11	17.5
Reduce labor	9	14.3
Cost savings	6	9.5
Upgrade	6	9.5
Productivity	5	7.8
Flexibility	5	7.8
Information requirements	5	7.8
Reliability	2	3.2
Reduce maintenance	1	1.6
Ease of maintenance	1	1.6
Eliminate labor	1	1.6
Safety	1	1.6

In a household-products company the interviewee commented that "in general it is harder to justify computer-based equipment because it is a systemwide integration which has other intangible benefits. It is hard to put a dollar value on things like improved quality and a lowered on-hand inventory level." Another remarked that cost reduction justification was "creatively done" in order to meet company requirements, but "it [cost reduction] is hard to find, when the primary goal of adoption is flexibility and faster turnaround time. It is hard to put dollar amounts on intangibles for accounting purposes. It is much easier to justify adopting new technology for new parts because there is nothing to compare costs with when we haven't built something before."

Formal recognition of the difficulty of justifying new technologies by traditional ROI measures, and of the importance of less tangible and quantifiable factors, has changed a few companies' justification procedures. Though a small change, it was a new policy of one company nevertheless to approve a project with a 25 percent ROI instead of the usual 30 percent if it would produce a quality improvement. A heavy-equipment manufacturer still uses cost reduction as the primary justification criterion, but because "direct labor is insignificant by now, we also look at indirect labor overhead, which is increasing." Another interviewee commented that "in the old system it was labor cost savings; in the new system it is reduction of any cost, such as inventory, overhead, and so forth."

A manufacturer of transportation equipment that had recently undergone restructuring to increase their competitiveness vis-à-vis Japanese companies, used multiple criteria: "cost/benefit analysis and ability to meet strategic goals such as market benefit, gaining on the competition, and flexibility gains." A large manufacturer of small durable goods, which had redesigned its manufacturing lines and procedures to improve quality as a primary strategy to compete with Japanese companies, had shifted its criteria for adoption and justification. The respondent told us of one instance in which the basis for adoption was "flexibility and providing real-time information on reliability to hourly workers, which helps productivity." Justification was in terms of "feasibility to meet on-time delivery and reliability. Of all the plants surveyed in our company, only one says they still use cost justification criteria."

These comments indicate there is widespread recognition of changes in both the technology and the terms of competitiveness that require different procedures and criteria for justifying new equipment. While responding to our questions, a number of interviewees volunteered comments indicating growing corporate awareness of the changes that were occurring, and that they were developing formal procedures to incorporate these intangible and harder-to-quantify factors in their justifications.

EXAMPLES OF COMPUTER-BASED PROCESS INSTALLATIONS

We questioned the companies regarding design and selection of programmable equipment for specific production processes. We asked each respondent to identify one or two recent instances in their company in which computer-based or programmable production equipment had been designed, acquired, or installed. We collected descriptive information about the type of process and equipment and whether the people involved were guided by or employed human factors policies on the project. We then asked each respondent to describe the outcomes of adopting this equipment. Performance outcomes were assessed in terms of production results after installation of the equipment (e.g., productivity, quality, etc.); the impact on workers was assessed in terms of skills, training, employment, and attitudes. During our interviews with 63 companies, 106 new equipment projects were described to us.

Types of Equipment

Nearly half of the processes given as examples were for mass production, 40 percent were for batch production, and 12 percent were for one-at-a-time manufacturing. The processes ranged from forming, cutting, and machining metal parts to finishing, assembling, and packaging final products (see Table 2–4). These processes involved all levels of computer-based automation. Nearly half of the processes included a minicomputer and/or a programmable controller, almost a third used microprocessors, 23 percent had central DNC, and 10 percent had numerical control. One-third of the systems were stand-alone machines and two-thirds were linked into an automated material handling system.

Selection, Design, and Implementation

The majority of systems (55 percent) were purchased from a vendor with little or no design input from the buyer. Twenty-eight percent were substantially designed in-house, and 17 percent were partially designed in-house (e.g., the system design but not the equipment design). In about a quarter of the processes the technology used was unique to the firm, 12 percent had technology that was specific to the industry, and almost two-thirds used technology that was also used in other industries and firms, and thus not specific to production of their particular product. Forty percent of

TABLE 2–4 Manufacturing Processes Involved*

	Frequency	Percent
Metal forming or cutting	37	35.6
Machining	36	34.6
Assembly	36	34.6
Testing or QC	27	26.0
Materials handling	25	24.0
Finishing	14	13.5
Packaging	5	4.8
Other	14	13.5

*Many examples included more than one process type.

the processes were designed and/or purchased and installed in under one year, 44 percent took between 13 and 24 months, and 15 percent took longer than two years. The mean project completion time was 17 months. Completion of the projects occurred on time in 51 percent of the cases, ahead of time in 6 percent, and delayed in 44 percent of the cases. For those delayed, the average time of delay was five months, and the median was three months.

Performance

Although performance of the new technology was generally superior to the technology it replaced, this is the outcome one would normally anticipate from any technological change. Productivity of the new systems was better than those they replaced in 93 percent of the cases. Flexibility and reliability were better in the new processes in three-quarters of the cases. The most significant benefit of the new processes over the old systems was in quality: 96 percent reported quality improvements (Table 2–5).

We also asked, however, how the new technology measured up to the designers' expectations for it. Here we found a somewhat less positive picture. Successes still outnumbered failures on all scales, to be sure, but, as Table 2–5 also shows, there were a significant number of instances where expectations were not met. Because respondents were given the opportunity to select any of their projects to report on, it is unlikely that they would have chosen any real design failures. The rather high incidence (approximately one in five) of lower-than-expected results, therefore, is noteworthy. For example, project delays of from 4 to 24 months were reported in almost half (46 percent) of the instances of late completion.

Impact on Work Force

The impact of the new technology on the work force was assessed in terms of employment levels, changes in skill levels, and training provided for operators, technicians, and mechanics. The processes were operated by union workers in 60 percent of the cases. In most companies, our respondents reported little negative impact, and a significant number reported a positive impact on operators, mechanics, and technicians working with the new equipment.

For the three job categories, skill levels and pay rates remained the same or increased in most of the instances. In 48 percent of the projects, operators remained at the same skill level, while in 41 percent they had a formal increase in skill level. Only 27 percent, however, also had a pay increase. As might be expected, mechanics and technicians had more instances of increases in skill, but even fewer had increases in pay rates. In nearly two-thirds of the new installations, mechanics and technicians had increases in skill, but only 10 percent of mechanics and 17 percent of technicians had increases in pay rates (see Table 2–6 and Figures 2–7 and 2–8). Virtually all companies reported the availability of classroom and/or on-the-job retraining for their workers. In only 3 to 4 percent of the instances did operators, mechanics, or technicians receive no retraining.

TABLE 2–5 New Technology Performance

	Compared to Previous Process		Compared to Expectations	
	Frequency	Percentage	Frequency	Percentage
Productivity				
worse	1	1.1	17	22.7
same	6	6.4	29	38.7
better	87	92.6	29	38.7
	94		75	
Flexibility				
worse	10	11.0	9	11.8
same	13	14.3	42	55.3
better	68	74.7	25	32.9
	91		76	
Quality				
worse	0	0.0	6	7.8
same	4	4.3	33	42.9
better	89	95.7	38	49.4
	93		77	
Reliability				
worse	2	4.4	14	20
same	10	22.2	35	50
better	33	73.3	21	30
	45		70	
Cost				
higher cost	7	13.7	15	21.7
same cost	10	19.6	35	50.7
lower cost	34	66.7	19	27.5
	51		69	
Downtime				
worse	11	14.3	15	22.1
same	26	33.8	35	51.5
better	40	51.9	18	26.5
	77		68	
Return on Investment				
worse	3	7.7	12	18.5
same	11	28.2	36	55.4
better	25	64.1	18	26.5
	39		66	
Project Completion				
delayed	44	43.6		
on time	51	50.5	(data not available)	
ahead	6	5.9		
	101			

The employment impact resulting directly from introduction of these new processes was difficult to assess. A substantial number (53 percent) of instances involved decreases in employment levels of operators. In contrast, a decline for mechanics occurred in only 10 percent of the processes and for technicians in 16 percent

TABLE 2–6 Changes in Worker Skill Level

	Frequency	Percentage
Change in operator skill		
less	10	11.0
same	44	48.4
more/higher	37	40.7
	91	
Change in mechanic skill		
less	1	1.3
same	28	35.4
more/higher	50	63.3
	79	
Change in technician skill		
less	2	2.7
same	26	35.6
more/higher	45	61.6
	73	

of the processes. In many of these cases, however, the employment change was not entirely due to productivity changes. Companies often reported that changes in their production processes occurred during a downturn in their product markets, so employment changes could also be attributed to declines in sales. The introduction of a new process, which improved quality and productivity, frequently allowed the company to stem further declines in sales, and was considered to have saved the jobs that remained. In nearly all cases, displaced workers were transferred to new jobs within the same company.

There were marked changes in operator attitudes toward the new processes before their introduction and after installation. In a quarter of the processes the interviewees reported that, prior to installation, the consensus view of operators toward the new system was negative, but in only 4 percent of the processes did this remain negative after installation (see Figure 2–9).

It is clear from these findings that computer-based technology does not necessarily reduce the need for worker skill. Such technology often requires a greater reliance on, and upgrading of, worker skills. As would be expected from new technology that was almost universally more productive than prior processes, some displacement of workers, primarily operators, did occur. The degree of displacement, however, was masked by volume changes, and the impact was moderated by the fact that in most of these companies workers were able to find other employment within the firm.

Many of the respondents felt that the new equipment would increase their companies' competitiveness and thus, ultimately, preserve jobs. Worker attitudes seem to support this view, because there was a decidedly positive shift of operator attitudes from prior to postinstallation of the technology. Investing in new technology was said by a number of companies to indicate to workers that management was committed to investing in the future of the company.

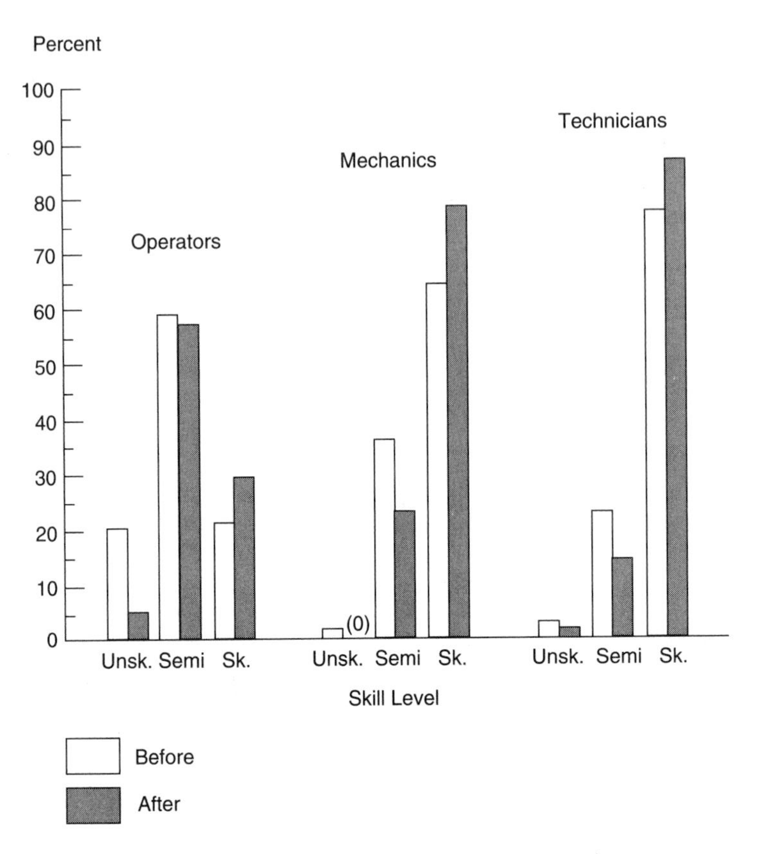

Figure 2–7 Skill levels before and after technological change

HUMAN FACTORS POLICIES

The central objective of the survey was to identify the types of human factors policies that companies were using to govern the design or selection of advanced manufacturing systems. We wished to compile a list of these policies, to identify design principles arising within these policy areas, and to gain insight into the impacts of such principles on the production process.

During the research it was important to distinguish between (1) policies that apply specifically *during* the design and selection stages of the acquisition of new manufacturing technology, and (2) general human resource policies relating to installed technology and the organization of work in the firm. The latter types of policies, which have been the major concern of the sociotechnical school of thought, have been the predominant focus of research on machine–human interaction. It was not our intent to put our plow into that field.

When we examined the evolution of literature on the design and development of manufacturing machinery (described in Chapter 3), we did not find a single source

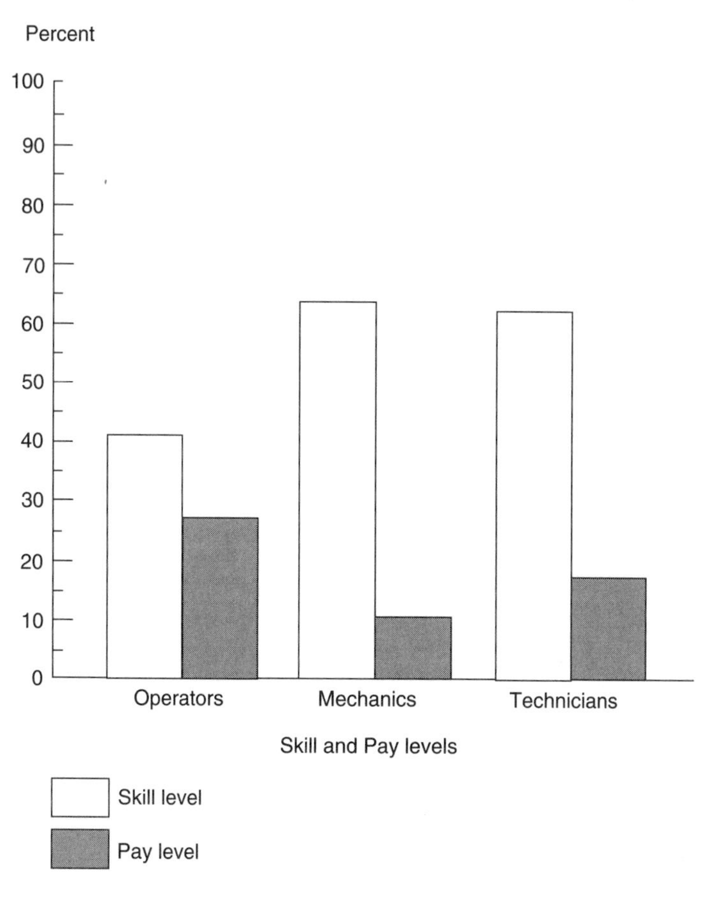

Figure 2–8 Increases in skill and pay levels

that provided a full set of human factors principles to guide designers. It has been our hypothesis that the most effective production, in terms of both quality and productivity, is accomplished through the design of manufacturing technology that builds upon human capabilities rather than excludes them. In such systems both the production process and worker development are enhanced. We sought, therefore, to go beyond the safety, health, and ergonomics policies that are fairly common in the design literature, to find those design and selection policies that recognize the full spectrum of human capabilities.

Policy Areas

We have identified a comprehensive set of policy areas pertaining to design and selection for machine–human compatibility. The remarkable feature of this list is its brevity. There are only 12 policy areas that need be the concern of the designer—

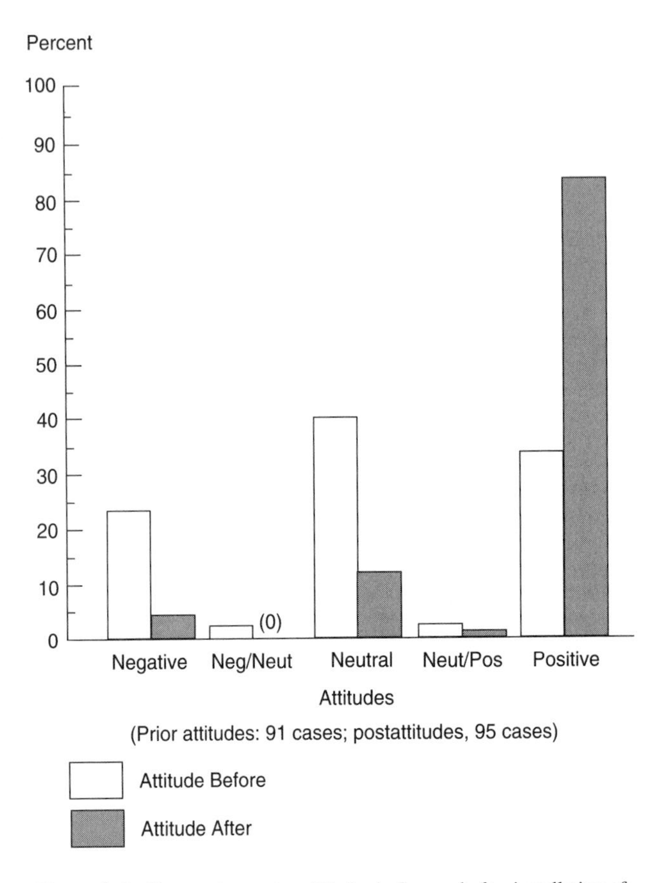

Figure 2–9 Reported operator attitudes before and after installation of new technology

merely a handful. This list is most helpful, both to managers who must set design policy and to educators who must teach designers to heed human factors principles.

The human factors policies we have identified fall into two categories: those that are concerned with the *physiological needs, limits, or capabilities* of people working with the technology, and those that are concerned with their *knowledge-based skills or capabilities.* The five physiological policy areas—safety, health, comfort, stress, and ergonomics—tend to be areas in which the policies are mandated by law or by corporate interest in avoiding liability. Technology design and selection policies in this category are intended to protect workers from adverse effects from the technology and to ensure physical consonance between machine and worker. Companies in our survey rather uniformly had design policies related to these areas. In our survey of design textbooks we found that these were also the human factors areas most frequently covered.

Cognitive capability policy areas, on the other hand, are far less frequently considered at the design or selection stage. These are the human factors areas that relate to how the worker's mental abilities are utilized, both during the design and selection stage and later when the new technology is in operation. The seven policy areas in this category are described below. In each case, we have defined the policy area in the form of a question or two to be answered by the people who are setting design policy.

1. *Involvement in the design and selection process.* Are the people who will be working with the new technology (operators, mechanics, technicians, and the like) or others who have relevant experience on current technology, expected to take part in various aspects of equipment design and selection?
2. *Operator control of the manufacturing process.* To what degree are operators to have control, that is, actual interaction with the process while it is in operation?
3. *Feedback of production information to the operator.* Is the new technology to be designed to provide operators with prompt data on process status, product quality, production output, yields, costs, and similar matters?
4. *Modification by the operator of equipment and process settings.* Is the process to be designed so the operator can vary the pace, override the machine sequence, modify the software, or in other ways adjust the process? Or is the system to be designed to prevent the operator from modifying the process?
5. *Worker skills considerations during design and selection.* Are there guideline policies on the level of skills desired relative to the new technology? Are machines to be designed to current skill levels in the firm, to higher levels, to lower levels?
6. *Maintenance and servicing of equipment by operators.* Is the operator to receive diagnostic information on machine condition, and is the operator to perform first-level maintenance tasks?
7. *Participation in postinstallation design changes.* Are workers to be encouraged to identify process problems and suggest changes?

Inherent in many of the knowledge-based capability policy areas is the issue of worker autonomy, or freedom to be actively involved in all aspects of the production process. As we had expected at the outset of the project, we found that relatively few of the firms had explicit policies in any of these areas. Even in firms that had practices that fell within one or more of the policy areas, there were few formal policies to guide the designers and buyers of new systems. A company might, for instance, routinely ask that operators participate in task groups to design or select new machines, but there would be no explicit policy that this should be the operating mode.

Survey Findings

The general conclusion that can be drawn from the survey is that even within the leading manufacturing companies in the United States, few have explicit policies that recognize the importance of taking advantage of the cognitive capabilities of the work force when new technology is being designed or selected. Companies provide rules to designers for dealing with the physiological requirements and limitations of the work force, but do not provide rules for involving peoples' broader capabilities. According to our survey, any company that had formal policies in this area was decidedly in the minority.

Table 2–7 summarizes the statistics on policy responses. In the "mandated" areas of safety, health, comfort, and stress, companies had equipment design and selection policies that were based largely on Occupational Safety and Health Act (OSHA) standards. Nine out of ten firms had explicit policies that met or exceeded OSHA standards for safety. Only one in ten had no formal safety policy with respect to equipment design or selection. Explicit policies for the other aspects of physiological concerns were less common. Sixty-eight percent of the responding firms had formal policies regarding health, 61 percent had policies regarding stress, and 51 percent had policies regarding comfort.

In two policy areas related to knowledge-based capabilities—involvement of workers in the design process and participation in postinstallation process changes—the policies ranged from informal worker consultation and involvement on an ad hoc, project-by-project basis to formal policies requiring worker membership on committees for the design and selection of new equipment.

Most companies did not have explicit policies regarding worker involvement in the design process, or for worker participation in modification of the process after it was installed. Thirty-three percent of the companies had informal procedures or opportunities for such worker input, but only 8 percent formally required worker involvement in equipment selection and/or design committees. Postinstallation worker participation occurred informally in 15 percent of the companies, but explicit policies calling for this participation were present in only 7 percent of the companies.

The question of skill considerations in design was asked as an open-ended inquiry about whether there was any policy requiring engineers to design equipment with respect to worker skills. Only 20 percent said that skill issues were a "consideration" and only 11 percent said that they had explicit policies relating to skills. Most of the considerations and policies were to lock out hourly workers and reduce labor force skill requirements. Only four companies (6 percent) reported using design policies that were oriented toward utilizing worker skill to improve production. Two of these companies tried to decrease worker interactions with the operation of machinery so workers could assume the broader role of using their analytic skills and judgment for process and/or system improvements.

Policies specifying equipment or system functionality were examined in three areas: feedback of production information to the operator, modification of the process by the operator, and maintenance by the operator. To implement modern quality-control methods, such as statistical process control, the operator needs information about the immediate production process. Yet, over two-thirds of the companies did not have an explicit policy regarding feedback of information to the operator; only a quarter of the companies had a policy to design equipment or processes to

TABLE 2–7 Summary of Policy Findings*

Physiological Requirements	Percent of Firms
No formal policy	12%
Meet OSHA standards	42%
Exceed OSHA standards	46%
Knowledge-based Policy Areas	
Worker Involvement in the Design Process	
No policy (no regular involvement)	59%
Informal requirement for involvement	33%
Formal requirement for involvement	8%
Design Equipment with Regard to Worker Modification of Process	
No policy	56%
Policy: lock out/prevent worker from modifying process	22%
Policy: design controls to allow/facilitate modification	22%
Information Feedback to Workers	
No policy	68%
Limited production information (SPC or QC)	25%
Provide general production information	7%
Worker Skills	
No policy	69%
Informal consideration of skills	20%
Formal policy: lock out/deskill	5%
Formal policy: utilize/develop worker skill	6%
Maintenance by Operators	
No policy	53%
Prevent operator maintenance	17%
Allow/facilitate operator maintenance	31%
Postinstallation Participation in Design Changes	
No policy (no regular participation)	78%
Informal requirement for participation	15%
Formal requirement for participation	7%

*Policies related to the design and/or selection of manufacturing technology within the firm.

provide production information for the purposes of statistical process control or quality control. Another 7 percent provided more general information about production.

We also asked whether they designed systems and equipment controls to provide operators with the capability of modifying or controlling the process. Twenty-two percent had policies for designs that allowed this type of operator control, while another 22 percent had policies specifying that the equipment be designed to lock out any type of operator intervention. Most of the remaining companies did not have any policy.

In terms of maintenance policies regarding operators, most companies (53 percent) did not have any policy, but 31 percent specified that equipment should be designed so that operators could perform first-level maintenance. Seventeen percent reported a design policy that operators would not perform maintenance.

POLICY DIVERSITY

The statistical aspects of the survey give a hint of the diversity of corporate attitudes toward the knowledge-based areas of human capability. The more qualitative responses we obtained relative to both informal and formal policies served to accentuate this diversity. In many instances policies reflected different company cultures, but there were also instances where the differences were clearly based on the nature of the process involved. Operator control or process modification was discouraged, for example, where there was high risk of harm to either operator or product.

Policies related to worker involvement in the selection and/or design of new equipment and in the postinstallation design changes were primarily on an informal basis. In a few companies the union contract called for involvement. At one company, we are told that "Once the functional specification is put together by manufacturing engineering, they will review it with the people who run the process. The engineers will preselect vendors to three good companies, and then the people who run the process participate in the review."

Several interviewees felt that worker participation was vital, and their companies required such participation. In the words of one interviewee, "Otherwise we don't get specifications that are needed, and the operators have a negative reaction. [Because of this participation] workers now propose new equipment needed."

Worker participation in changes made after equipment installation was more frequent, but generally informal. Many companies had suggestion systems, but in practice there was little active participation by workers and minimal follow-up by manufacturing engineers. Postinstallation participation appeared to be most effective in companies that required some type of formal participation system or had other types of programs that included workers. For example, companies using statistical process control would have various types of quality circles that would address equipment and process design as one area to help in quality improvement. Generally, however, there was little recognition of the value of worker involvement either before or after design and installation.

Maintenance by operators was often minimized because it was viewed by management, or mandated by union contract, as a higher-skilled activity that operators should not be allowed to do. In a few companies, however, workers were encouraged to perform routine maintenance on the equipment with which they worked. Encouraging operators to maintain their own machines tended to be a new policy that had been implemented with the goal of reducing the number of job classifications or of increasing worker "ownership" of the process. This approach was often a job policy rather than a design policy. In only a few of the companies were there attempts (but not policies) to provide diagnostic information and/or equipment designs with greater access in order to facilitate operator maintenance. Equipment designers at one company were beginning to design "smart panels" that self-diagnosed equipment problems and issued voice commands to guide the operator in fixing the problem. Another company was trying to design equipment with modules that could be replaced by the operator. The company's reason for doing this was to minimize dispute-related maintenance issues among certain trades, such as electrical and plumbing, because replacement of modules, as distinct from repair, was not a procedure under the jurisdiction of a given trade.

The design approaches in terms of worker skills were consistent with the approaches in the areas of control and maintenance. Typically, comments about skill (even when there was no policy) were to "simplify the machines to lower the skill required of our labor force"; to "automate to remove worker influence in order to improve consistency"; to "centralize control, so there is no operator modification of the process"; to "lock out the operator so he only *monitors* speeds and feeds"; and to ensure that the "programming unit has control of the machines, with a policy to make the process as independent of people as possible." The ideal in design was stated by one manager as "automating as much as possible—if you have an operator interacting with the process to keep it going, it causes problems—[and when there are problems] you call in skilled people to maintain the system. Keep it automatic without direct labor interacting."

A distinct minority of companies took design approaches that relied on the use of worker skills, albeit a different set of skills than previously. One company used automation to "eliminate the boring repetitive, unsafe jobs so we can get broader worker involvement and skills," and another also tried to decrease the amount of "manual dexterity [required in a process] to increase the judgment-based work, using statistics and understanding the information on the computer screens." One company relied on skill upgrading of their labor force and designed machines for greater local control of the processes; another designed "for fewer and more skilled operators," designing complex manufacturing and assembly equipment to yield the greatest productivity improvements possible, with the recognition that this would require much greater skill on the part of operators.

Companies in the survey also took divergent approaches to improving product quality, process reliability, and efficiency. The predominant approach to process design was from the perspective that full automation entails locking out or eliminating direct human intervention in the production process. At one company, a manufacturing engineer stated that they use "standard IE [industrial engineering] operating policies: keep the operator away from the controls. The machine controls quality and the operator only loads and unloads the machine." Companies using this approach typically had policies to the effect that programming was to be done by a salaried group of technicians and not by operators.

Operator control was viewed as introducing elements of uncertainty into the process. As expressed by one manufacturing engineer in a company where operators did not have access to the controls, "If the knobs are available to the operator, he will do what he wants." This would vary the process in nonstandard and unpredictable ways, and "bastardize the process," in the words of another manager. One interviewee said, "Operators are just that and rarely program; we discourage operator control and want central control." Another manager viewed process control as a labor-management issue, stating a policy of "removing human control lets people be parts feeders or material handlers where they cannot affect outcomes. We won't allow the union to shut us down!"

In other companies, the human dimension of production was viewed quite differently. Operators were regarded as "owners" or "managers" of the production process with which they worked. From this perspective, feedback of production information to the operator and control or modification of the process by the operator were considered appropriate policies. Companies that designed equipment or systems to provide operators with production information were generally using statistical process control (SPC) or other quality procedures. One company was implementing statistical process control throughout the company and designing and retrofitting workstations with computer terminals that provided operators with necessary SPC data. Instances when operators overrode automatic controls were carefully regulated, however, and had to be reported to their supervisor.

In companies such as these, operator involvement with information could be either passive or active. In a passive mode, the operators monitored and recorded production activity and passed problems on to technicians. In an active mode, information was provided so the operator could take corrective action if needed. In a few of the more advanced companies, the equipment was designed to facilitate operator modification of the process.

Companies reported significant benefits from providing SPC information to operators. One company stated that it allowed "real-time verification of specifications, finding defects, and results in first-time quality with no rework."

In another set of companies—a distinct minority—operators were expected to participate actively in the production process, and information was provided to assist them. The production systems were often referred to as "operator-owned" or "operator-managed" processes. "Operators are trained to be self-sufficient and to work generally without supervision to do what is necessary, including adjustments, if the job calls for it," was the stated policy in a company supplying automobile parts. Expressions of this perspective included statements such as, "The system is the operator's; each is responsible for the operation of the system," or "The operator should almost always be permitted to override; each individual operator is responsible for quality, and parts are traced," or "we emphasize developing designs which allow the operator to modify controls, once sufficient information is available to know how to do it effectively, and that gives operators SPC information on-line."

A few companies were actively training their operators and letting them run the production process with minimal supervision. One manufacturer of system controls trained operators in problem solving and expected them to apply their training, when needed, "within specified boundaries." In perhaps the most far-reaching policy of this nature, one company designed a variable speed assembly line that was directly controlled by an operator chosen each week by the entire group of operators. There was an incentive pay system, and in the first week after start-up they were running the line at over 110 percent of the base rate. In a different area of this same company, operators were given a recommended method of operation but were expected to modify it as they found necessary.

Thus, we found that the companies took widely divergent approaches in their quest for production process and product improvement. The predominant approach was to use new technology to attempt greater automation without intervention or control by operators, pushing decision making up the job ladder to technicians, programmers, and supervisors. At the other end of the spectrum, however, were a few companies that in varying degrees were assigning to operators a large share of responsibility for the production process.

SUMMARY

We expected that large manufacturing firms who were leaders in highly competitive product markets would have the incentives and the resources to find practices and policies to keep them ahead. The survey confirmed this expectation. These firms, we found, are spending large sums on new manufacturing technology. New approaches to process design are also emerging, but few policies with respect to human involvement with the new technology have been formalized.

The examples of new technology adopted indicate that the companies were largely successful in achieving the objective of increasing machine performance. Increasing human performance was also involved, as was indicated by increases in skills required to operate and maintain the new systems. It is clear, however, both from the evidence of lags in pay changes relative to skills increases, and from the lack of specific policies regarding skills, that a large majority of companies had not considered the matter of skills when the equipment was designed or selected.

In fact, our survey quite definitely establishes that even market-leading firms have paid little or no attention to human factors (other than the mandated safety and health issues) in the acquisition of new technology. Several reasons for this situation can be advanced. The most obvious is the historic distance and tensions between management and the work force. But this research suggests other reasons as well. The self-rating of the companies according to people and task concerns indicated that firms are considerably more focused on getting the work done than they are on the people who are doing the work. Further, the evidence from engineering textbooks indicates that engineers and designers are highly unlikely to be exposed to any comprehensive coverage of human factors issues in their training. They are likely to be predisposed to ignore such considerations as being outside the engineering task.

The companies in our survey were not, however, entirely devoid of ideas or policies on how to incorporate knowledge-based factors into equipment design. We found that many had informal policies or practices that could be identified and categorized. Upon completing the survey, we found we had a concise, but comprehensive, list of possible human factors policy areas. This list (Table 2–8) provided us

TABLE 2–8 Human Factors in Manufacturing System Design

Physiological Capabilities/Limitations
Safety
Health
Comfort
Stress
Ergonomics
Knowledge-based Capabilities
Involvement in the design process
Operator control of manufacturing process
Feedback of production information
Modification of equipment or process settings
Skills utilization
Maintenance responsibility
Participation in postinstallation design changes

with a framework within which to conduct the second major stage of our research. This stage consisted of detailed field studies of the design of major manufacturing systems in five large companies. The research procedures and the case studies themselves are presented in chapters 5 through 9.

These policy areas, in turn, guided our search for principles that should apply during the design and selection of new technology. The fruits of that search are found in the final chapter of this book.

3

HUMAN FACTORS DESIGN PRINCIPLES: A TEXT REVIEW

One reason for conducting this project was to define human factors principles that could be taught to engineers who would be designing manufacturing systems. It was our conviction, based on experience in teaching and management, that an engineer's education is woefully lacking with respect to human capabilities, needs, and limitations. To demonstrate this, we decided to examine what engineers have been and are being taught on this subject.

We went to engineering textbooks and handbooks to see what the experts in engineering design and machine design had to say. This part of the study was done in parallel with the survey described in Chapter 2, so we had a fairly good idea of what topics to look for. What we found was an evolving trend toward greater mention of human factors in production systems, but even the latest textbooks treat human factors from a point of view that differs little from the scientific management theories of the early 1900s.

We reviewed over 100 books on machine systems design and on engineering of production systems in the Boston University Science and Engineering Library and 100 textbooks used in engineering courses. The review was done in two steps. The first involved an examination of 25 engineering handbooks published over a 50-year span from 1938 to 1987. The intent was to gain a historical perspective of the treatment of human factors topics over this time period. The second review broadened the search to include textbooks used in a variety of engineering courses. In all, over 200 books were reviewed. Out of the combined efforts of these two approaches, only 42 books were found to have any mention of process design issues involving humans.

HANDBOOK REVIEW

In the historical-perspective study, 25 engineering and machine design texts published between 1938 and 1987 were examined for human factors content. In terms of mentioning a given topic, the trend toward increasing awareness of human factors is clear, but not consistent. Figure 3–1 portrays graphically the extent to which various human factors topics are mentioned by each textbook.

One observation made while conducting this study was that recent texts tended to refer the reader to an ergonomics handbook or an industrial engineering handbook, rather than cover human factors topics in the text itself.

DESIGN TEXT REVIEW

When we examined the balance of the 200 books in the review, we searched their tables of contents and indexes for any listing of human factors, ergonomics, safety, skill, or names of any management theorists such as Frederick Taylor or Douglas McGregor. The 42 books we found having some mention of human factors are listed among the references at the end of the chapter.

Predictably, in these 42 books there is marginal consideration of the role of people as part of a production system. "Important" or "primary" considerations involve a list of factors about machine performance (e.g., cycle times, throughput). Design-decision criteria are generally based on variations of "economy and efficiency" themes. Virtually absent is discussion of facilitating active operator interaction to promote systems performance. Following are excerpts of typical discussions of human considerations in design that illustrate the authors' implicit and explicit conceptions of design considerations relative to the people working with the machines.

An early book that discusses systems philosophy (Ellis and Ludwig, 1962) notes that "The optimal performance of the element 'man' in man–machine control system performance can be obtained only when the mechanical components of the system are designed so that the human being need only act as a simple amplifier" (p. 79). A more recent formulation of systems-engineering principles discusses personnel objectives as being to eliminate people in systems. When, however, they are required, "skill level requirements should normally be minimized" and designed for workers who are "able to follow clearly presented instructions where interpretation and decision making are not necessary [and] will normally require close supervision" (Blanchard and Fabrycky, 1981, pp. 435 and 436). Another author begins a section entitled "Human Factors" with the reminder that "The engineer must never forget that whatever he designs is meant to be used by human beings. . . . unless the designer goes to apparently absurd lengths to prevent it, people will operate the device incorrectly" (Gibson, 1968, p. 23).

In the few books that do discuss explicit design principles relative to workers,

REF. NO.	01	02	03	04	05	06	07	08	09	10	11	12	13	14	15	16	17	18	19	20	21	22	23	24	25
PUBLICATION YR	'38	'43	'51	'55	'55	'56	'57	'57	'59	'61	'62	'63	'64	'65	'66	'68	'69	'70	'75	'82	'83	'84	'85	'86	'87
DESIGN TYPE	M	M	M	M	M	M	M	M	M	M	E	M	M	M	E	E	E	E	M	M	M	E	M	M	M
HUMAN FACTORS																									
SAFETY	●	●									●		●		●	●	●	●	●		●	●	●	●	●
ERGONOMICS											●		●		●	●	●	●	●		●	●	●	●	●
CONTROL													●			●								●	●
MODIFICATION																●		●							
FEEDBACK															●								●	●	●
PARTICIPATION											●					●						●			
MAINTENANCE		●																●			●			●	●
DESIGN CHANGE																●									
SKILL																								●	●

● - Mentioned as criterion　　M - Machine design　　E - Engineering design

Figure 3–1 Study of design texts

the most frequently cited reference is Frederick Taylor, followed by the Gilbreths. These discussions cover the essentials of time-and-motion studies. Typical statements: "in order to increase productivity, a worker should be paid per unit of production rather than on an hourly basis. . . . [this] is called time study" and, "quite often a scientifically trained engineer can study the way a worker goes about his job and suggest a more convenient arrangement of work space and a more efficient series of motions of the worker. . . . [this] is called motion study" (Gibson, 1968, p. 164). Translating time-and-motion-study principles into design principles involved generally broad precepts such as "minimize the total number of motions required," or "work should be distributed as equally as possible over the two hands and two feet."

Human factors considerations, along with time-and-motion principles, are illustrated in an example of designing a workbench for installation of electrical resistors. This assembly process required that after determining the "good" and the "bad" resistors, the worker would deposit them in different bins. To design the workbench:

> Paint the good hole edge [of the bin] green, the bad hole edge red. The girl should not read a meter. She should not be told what an ohm is. She should not be asked to make difficult borderline judgments. A green light should go on if the resistor is good and a red light if it is bad. . . . The foreman should make the setup and test it before turning it over to the girl. Any adjusting of the test procedure may upset the girl, and any slight gain in efficiency will be lost by retraining time.
>
> Note that the job of converting the actual quantity under test to a simple binary decision is done by the industrial engineer in designing the test circuit, not by the girl. . . . Theoretically it would be possible to let the girl watch the dial on an ohmmeter [and decide if the resistor were good]. For the highly trained technician in a laboratory this is perfectly satisfactory, but not for girls on an assembly line for whom the procedure should be wholly automatic. (Gibson, 1968, pp. 166–167).

The authors of a book on systems design conclude their chapter entitled "Human Subsystems" with the following summary:

> Having indicated some of the problems associated with human subsystems . . . and having compared certain characteristics of men and machines, we may well raise the question of why human subsystems should be utilized at all in systems design. [The reason is partly because of the limitations in fully automating systems to run without any human intervention, and,] more practically, as has been humorously pointed out by many, man represents a fairly high capability control element, already inexpensive in mass production and producible by inexperienced labor. (Ellis and Ludwig, 1962, p. 81).

More specific considerations of workers during machine design are presented in terms of safety and traditional ergonomic factors. By the late 1960s, safety becomes a consistently mentioned item in design books, presumably as a result of Occupational Safety and Health Act standards. (OSHA became law in 1970.) Safety requirements introduced design considerations for humans as a factor of production apart from equipment performance. This marked a shift in conventional machine design thinking, because safety involves design for the benefit of the human operator, not necessarily machine performance. Brief lists of safety considerations appear in most books written after the mid-1960s, and many advise the designer to consult the applicable safety codes of his or her state, industry, or professional association. Most discussions are short, reminding the designer that safety should be considered along with the "primary" criteria for good machine design.

Textbook presentations on ergonomics focus on the physical capabilities and limits of humans and the necessary design considerations of these factors to obtain optimal machine performance. Typically the sections on designs for machine–human interfaces discuss designing for the 95th percentile of height, reach, and so forth. Many of the books are replete with charts detailing average lift capacity, grip strength, reach, comfortable seat height, and reaction times to sound, sight, smell, touch and temperature. Most texts in recent decades tend to derive their material on ergonomics from measurement approaches detailed by Dreyfuss (1960). One text notes that understanding the limits of human capacities is important because, "requirements outside of these recommended limits will result in operator inefficiency and system failures" (Blanchard and Fabrycky, 1981, p. 427).

An example of the importance of designing for human capacities is the effect of noise on performance. Citing studies of steady versus intermittent noise which indicate that steady noise is less distracting than intermittent noise, even at high levels, the text states that thus "an individual can adapt to it and work efficiency may not be significantly compromised" (Blanchard and Fabrycky, 1981, p. 430). The authors do note that "if the noise level is too high (even though steady), the individual will probably experience permanent injury through loss of hearing," but, "on the other hand," intermittent noise, although less likely to cause injury, is distracting with loss of job efficiency, and thus the "noise generated by the system must be maintained at a level where human efficiency is maximized" (p. 431).

One other author, Phadke (1989), mentions operator error as one of the "noise factors" in engineering design, but there is no other mention of human factors that apply to design.

In the introductory chapter of their book, Juvinall and Marshek (1991) describe Abraham Maslow's hierarchy of human values as one aspect of what they call "societal values and considerations," an unusual departure for an engineering text. Designing for safety is also discussed briefly in the introduction to this book.

Another recent textbook (Niebel, Draper, and Wysk, 1989) has this lone reference to human factors:

> Under incentive wage payment plans the operator is usually paid for the good pieces he makes. In some organizations, the operator's pay is determined according to the inspector's records at the end of the day or at the completion of the order . . . Therefore, the responsibility of the quality-control man in approving an operator's pay is very important. (pp. 894–895)

Other recent textbooks (El Wakil, 1989; Shigley and Mischke, 1989; Suh, 1990) are devoid of any mention of human aspects in process design.

Consideration of any human factors, including safety, in any of these textbooks we reviewed is generally limited to one- or two-paragraph overviews, and these typically cover ergonomic concepts. One book concludes a brief section on human factors by noting that although the "numerosity of designs requiring consideration of human factors is extensive," the "competent designer handles these with little but careful deliberation and judgment" (Vidosic, 1968, p. 114).

In general, when human factors need to be considered in greater depth, most books advise the designer to consult the human factors specialist much as they would consult any other engineering specialty for expertise on a part or material with which they are not familiar. Human factors engineers constitute only 0.2 percent of all engineers, however, so they are not likely to be involved in a typical design project (Alluisi, 1987).

EVALUATION

This sampling of engineering texts and handbooks reveals an implicit concept of a worker as limited in ability to contribute to production performance. It is not the drive for automation, per se, that degrades the role of workers in production, but rather a negative perspective of worker capabilities that requires designers to reduce the risk humans pose to successful operation of the technology. By instructing the designer to reduce human action to "simple amplification" of machine operations, the authors apparently intend that humans should perform instructions prescribed by the machine rather than take independent action or analysis. There is no consideration that the reduction of task complexity might allow the operator to be engaged in other activity that would enhance system performance.

Interestingly, the authors of one design book (Blanchard and Fabrycky, 1981) suggest that the more limited the task complexity the greater the need for supervision, implying a lack of capability at the most basic level of worker functioning and/or a labor-management issue of worker unwillingness to engage in such activity.

The underlying premises of limited worker capabilities and worker recalcitrance or resistance are also present in the description of the time study in which the "scientifically trained engineer" is the person able and willing to suggest improvements in work efficiency. We see repeated in text after text that these assumptions regarding human capability and behavior are translated into engineering principles without mention of the assumptions or examination of their validity.

In the example of designing a workbench, the author demonstrates implementation of principles he has listed as "scientifically" determined for optimal performance. Although making a mundane task simple and rote repetition may be appropriate strategies for effective performance of that task, the design instruction is coupled with an implied concept of worker capabilities. The goal of making the task as automatic as possible does not necessarily require that the worker deliberately be kept ignorant about the task. It could be argued that additional knowledge would enable the "girl" to make suggestions for increased efficiency, conduct the setup herself instead of relying on a foreman (also increasing efficiency), and having an ohmmeter in addition to a light so she can provide feedback on the magnitude of error and margin of quality in the resistors. Moreover, there is a not-too-subtle message about the capabilities of women—they are easily upset and have low ability for retraining or interpreting ohmmeter readings.

The general presumption that workers provide only flexibly guided physical exertion or manipulation for operations difficult to automate is also evident in the prescription for work to be distributed equally over the hands and feet. These appendages are, apparently, the only aspect of worker capabilities relevant for conducting work. Given the limited attention devoted to ergonomics and other human factors subjects in the engineering texts, one is forced to conclude that these subjects are definitely peripheral to, not part of, the essential tasks of design.

In our literature review we find no appreciable change over time in the authors' views of the human role in production or in its importance as a design criterion. The more recent books tend to mention human factors or the "machine–human interface" to a greater extent than the earlier books. However, there is still quite limited discussion: A 1983 machine design book of 642 pages is typical. It devotes two paragraphs to human factors engineering, stating, "human factors engineering is concerned with all aspects of the man–machine relationship," and then listing all those aspects as "safety, comfort, and efficiency" (Hindhede, 1983, p. 22).

On the basis of this review, it would appear that humans are not viewed by the writers of machine design texts as having any significant potential contribution to production systems. Rather, they are to be considered as imposing external design constraints ergonomically or legally (safety and health). In general, workers are considered a "risk" to production systems, and the risk should be reduced by eliminating human intervention if possible, or by minimizing it to the greatest degree possible.

These principles and applications are the familiar methods of scientific management. What is noteworthy is that although management theory has been significantly amended since World War II, engineering design principles have not undergone a similar development. Engineering design, as indicated in these books, has proceeded into the current era with the same basic principles about the human role in production as were first articulated at the beginning of the twentieth century.

OTHER CRITIQUES

Our findings in this review are consistent with other researchers' studies of engineering objectives. David Noble suggests that the values of the engineering profession reflect those of management, because engineering has developed within the confines of industry and is thus subservient to management, rather than developing as an independent profession. He contends that these values reflect a set of intrinsic engineering goals. He finds among engineers "a delight in remote control and an enchantment with the notion of machines without men . . . a general devaluation of human skills and a distrust of human workers and an ongoing effort to eliminate both" (Noble, 1984, p. 191). He argues that these objectives dominated design choices in development of numerically controlled machine tools.

Perrow's (1983) study of human factors engineering finds that even standard human factors considerations are marginal in engineering design. He attributes the neglect of principles of human factors to "top management goals and perspectives, the reward structure of the organization, insulation of design engineers from the consequences of their decision, and some aspects of organizational culture" (pp. 523–524). In essence, the organization of engineering practice leads to lack of accountability for human factors, or for the consequences of designs that are not engineered with regard to human factors, and the organizational isolation of human factors engineers.

Although human factors engineering does focus on the "human subsystem," it is based on a rational, biological, and mechanical view of humans, according to Perrow. Human factors engineers have a view of the human as an information-processing system that responds predictably to positive and negative sanctions, in a manner similar to the mechanical view of the engineer. This shared view is of workers as "transfer devices" in the automation loop, who are used "for want of a robot." Thus, he argues, human factors engineering does not examine the social consequences of design choices.

In an extensive review of technology design models, Blackler and Brown (1986) note that ergonomics, sociotechnical theory, and the theory of participative management have had limited impact on design because the "climate of opinion" is not receptive to such theory. To influence technology design, they suggest greater "intervention" in design by social scientists, and psychologists in particular, to "help different groups within organizations to recognize and implement new approaches to technological innovation" to improve its social impacts (p. 308). Mumford (1987), in a historical review of sociotechnical theory, notes that very little of the research in this stream addresses technology design and suggests one reason may be the lack of engineering expertise by sociotechnical researchers.

CONCLUSION

Our examination of the tools of the trade in engineering education confirms that there is substantial room for revision and expansion of the treatment of the human contributor to productive systems. If, as a result of this study, we can identify the principles that need to be taught, an opportunity exists for future writers of machine design texts to depart from outmoded, conventional thought and to create truly modern texts.

REFERENCES

ALLUISI, EARL A. (1987). "The Human Factors Technologies—Past Promises, Future Issues" in L.S. Mark, J.S. Warm, and R.L. Huston (eds.), *Ergonomics and Human Factors: Recent Research.* New York: Springer-Verlag.

ASIMOW, MORRIS (1962). *Introduction to Design.* Englewood Cliffs, NJ: Prentice-Hall.

AULT, NORMAN (1938). *Fundamentals of Machine Design.* New York: Macmillan.

BAILEY, ROBERT W. (1982). *Human Performance Engineering: A Guide for System Designers.* Englewood Cliffs, NJ: Prentice-Hall.

BLACK, PAUL H. (1955). *Machine Design.* New York: McGraw Hill.

BLACKLER, FRANK, AND COLIN BROWN (1986, January). "Alternative Models to Guide the Design and Introduction of the New Information Technologies into Work Organizations." *Journal of Occupational Psychology,* pp. 287–313.

BLANCHARD, BENJAMIN, AND WOLTER FABRYCKY (1981). *Systems Engineering and Analysis.* Englewood Cliffs, NJ: Prentice-Hall.

BLUMENTHAL, MARJORY, AND JIM DRAY (1985, January). "The Automated Factory: Vision and Reality." *Technology Review,* pp. 29–37.

BRADFORD, LOUIS J., AND PAUL B. EATON (1963). *Machine Design.* New York: John Wiley & Sons.

BRÖDNER, PETER (1986)."Skill-Based Manufacturing vs. 'Unmanned Factory'—Which is Superior?" *International Journal of Industrial Ergonomics,* vol. 1, pp. 149–153.

CHERNS, ALBERT (1987)."Principles of Sociotechnical Design Revisited." *Human Relations,* vol. 40, (3), pp. 153–162.

DEUTSCHMAN, AARON D., WALTER J. MICHAELS, AND CHARLES E. WILSON (1975). *Machine Design, Theory and Practice.* New York: Macmillan.

DREYFUSS, HENRY (1960). *The Measure of Man: Human Factors in Design.* New York: Whitney Library of Design.

ELLIS, DAVID O., AND FRED J. LUDWIG (1962). *Systems Philosophy.* Englewood Cliffs, NJ: Prentice-Hall.

EL WAKIL, SHERIF D. (1989). *Processes and Design for Manufacturing.* Englewood Cliffs, NJ: Prentice-Hall.

ERDMAN, ARTHUR G., AND GEORGE N. SANDOR (1984). *Mechanism Design: Analysis and Design.* Englewood Cliffs, NJ: Prentice-Hall.

FAIRES, VIRGIL MORING (1957). *Design of Machine Elements.* New York: Macmillan.

GIBSON, JOHN E. (1968). *Introduction to Engineering Design.* New York: Holt, Rinehart and Winston.

GOSLING, W. (1962). *The Design of Engineering Systems.* New York: John Wiley & Sons.

GROOVER, M. P. (1987). *Automation, Production Systems, and Computer-Integrated Manufacturing.* Englewood Cliffs, NJ: Prentice-Hall.

HALL, ALLEN S., ALFRED R. HOLOWENKO, AND HERMAN G. LAUGHLIN (1961). *Theory and Problems of Machine Design.* New York: McGraw Hill.

HARRISBERGER, LEE (1982). *Engineermanship: The Doing of Engineering Design.* Belmont, CA: Wadsworth.

HILL, PERCY H. (1970). *The Science of Engineering Design.* Troy, MO: Holt, Rinehart and Winston.

HINDHEDE, UFFE (1983). *Machine Design Fundamentals—A Practical Approach.* New York: John Wiley & Sons.

HINKLE, ROLLAND T. (1957). *Design of Machines.* Englewood Cliffs, NJ: Prentice-Hall.

HIRSCHHORN, LARRY (1984). *Beyond Mechanization.* Cambridge, MA: MIT Press.

HYLAND, P. H. AND J. B. KOMMERS (1943). *Machine Design.* New York: McGraw Hill.

KALPAKJIAN, SEROPE (1989). *Manufacturing Engineering and Technology.* Reading, MA: Addison-Wesley.

JAIKUMAR, RAMCHANDRAN (1986, November–December)."Postindustrial Manufacturing." *Harvard Business Review,* pp. 69–76.

JUVINALL, ROBERT C., AND KURT M. MARSHEK (1991). *Fundamentals of Machine Component Design,* 2nd ed. New York: John Wiley & Sons.

MANO, M. MORRIS (1988). *Computer Engineering: Hardware Design.* Englewood Cliffs, NJ: Prentice Hall.

MOLIAN, S. (1982). *Mechanism Design.* New York: Cambridge University Press.

MUMFORD, ENID (1987). "Sociotechnical System Design; Evolving Theory and Practice," in Gro Bjerknes, Pelle Ehn, and Mortan Kyng (eds.), *Computers and Democracy—A Scandinavian Challenge.* Aldershop, England: Avebury Academic Publishing Group.

NICOLAI, LELAND M. (1975). *Fundamentals of Aircraft Design.* Dayton, OH: School of Engineering, University of Dayton.

NIEBEL, BENJAMIN W., ALAN B. DRAPER, AND RICHARD A. WYSK (1989). *Modern Manufacturing Process Engineering.* New York: McGraw-Hill.

NOBLE, DAVID F. (1984). *Forces of Production: A Social History of Industrial Automation.* New York: Oxford University Press.

PAHL, G., AND W. BEITZ. KEN WALLACE (ed.), (1984). *Engineering Design.* New York: Springer-Verlag.

PERROW, CHARLES (1983, December). "The Organizational Context of Human Factors Engineering." *Administrative Science Quarterly,* pp. 521–541.

PHADKE, M.S. (1989). *Quality Engineering Using Robust Design.* Englewood Cliffs, NJ: Prentice-Hall.

PRESSMAN, ROGER S. (1987). *Software Engineering, A Practitioner's Approach.* New York: McGraw-Hill.

ROSENTHAL, EMANUEL (1955). *Elements of Machine Design.* New York: McGraw Hill.

ROSENTHAL, STEPHEN, AND HAROLD SALZMAN (1993 in press). *Computing and Organization: Studies in Values and Software Design.* New York: Oxford University Press.

ROUSE, WILLIAM B., AND WILLIAM J. CODY (1987). On the Design of Man-Machine Systems: Principles, Practices, and Prospects. *IFAC World Congress,* München.

SALZMAN, HAROLD (1989, November). "Computer-Aided Design: Limitations in Automating Design and Drafting." *IEEE Transactions on Engineering Management,* vol. 36, p. 4.

SHARMA, P.C., AND D.K. AGGARWAL (1985). *Text Book of Machine Design.* Ludhiana, India: Katson Publishing House.

SHIGLEY, JOSEPH E. (1956). *Machine Design.* New York: McGraw Hill.

SHIGLEY, J.E., AND C.R. MISCHKE (1986). *Standard Handbook of Machine Design.* New York: McGraw Hill.

SHIGLEY, J.E., AND C.R. MISCHEKE (1989). *Mechanical Engineering Design,* 5th ed. New York: McGraw Hill.

SIEGEL, MARTIN J., VLADIMIR L. MALEEV, AND JAMES B. HARTMAN (1965). *Mechanical Design of Machines.* Scranton, PA: International Textbook Co.

SLAYMAKER, R.R. (1959). *Mechanical Design and Analysis.* New York: John Wiley & Sons.

SUH, N.P. (1990). *The Principles of Design.* New York: Oxford University Press.

TAYLOR, J.R. (1989). *Quality Control Systems.* New York: McGraw-Hill.

TOLLY, S.D., AND A. VAN DAM (1984). *Fundamentals of Interactive Computer Graphics.* Reading, MA: Addison-Wesley.

TORENBEEK, EGBERT (1976). *Synthesis of Subsonic Airplane Design.* Delft, Netherlands: Delft University Press.

VALLANCE, ALEX, AND VENTON LEVY DOUGHTIE (1951). *Design of Machine Members.* New York: McGraw Hill.

VALLANCE, ALEX, AND VENTON LEVY DOUGHTIE (1964). *Design of Machine Members,* 4th ed. New York: McGraw Hill.

VIDOSIC, JOSEPH P. (1968). *Elements of Design Engineering.* New York: The Ronald Press Company.

WALTON, RICHARD, AND GERALD SUSMAN (1987, March–April). "People Policies for New Machines." *Harvard Business Review.*

WILCOX, ALAN D. (1990). *Engineering Design for Electrical Engineers.* Englewood Cliffs, NJ: Prentice Hall.

WOODSON, THOMAS T. (1966). *Introduction to Engineering Design.* New York: McGraw Hill.

4

CASE STUDIES

PURPOSE

The second major stage of the research project involved a series of case studies of actual instances of design or selection of advanced manufacturing systems within companies. The purpose of this approach was to identify human factors principles that had actually been employed by leading firms and to evaluate the results that had been achieved through the use of these principles.

This was a pragmatic undertaking—we were not seeking to invent new principles, but to find those that had been shown to be effective in actual application. Experience has shown that there is very little in the way of human resource policy that is totally new. Most ideas have links to the past. Today's concept of multiskilling, for example, has direct relationships to yesterday's apprenticeship programs, and these, in turn, can be said to have roots all the way back to the guilds of the Middle Ages. Likewise, the Quality Circle approaches of the 1970s echo much of the work simplification methods of Alan Mogenson and others in the 1930s.

APPROACH

Each case comes from a different company. Each represents a detailed exploration of events and the policies guiding them in an instance of design or selection of manufacturing technology controlled by computers or computer-like devices. Case data were collected in the field—in the plants and offices of the companies studied.

Case studies as a research tool have both benefits and limitations with respect to what can and what cannot be determined. The benefits include the fact that an intensive examination of one situation at one time in the history of a company can reveal a great deal more about that situation than could ever be elicited by questionnaire or other research approach. Case studies give the researcher the opportunity to discover the unexpected, and to find linkages between factors that influence the outcome. Case studies permit the researcher to "triangulate" on the data—to obtain verification of information by querying multiple sources, including review of relevant documents. Teams that engage in case studies also become known to the subjects of the study and are able to establish a level of trust and responsiveness that is very difficult to achieve by survey.

Those who are engaged in case studies must also be aware of the limitations of such research. Because case studies place great demands on the human and financial resources of the research team, the number of cases that can economically be attempted is small. In this project, we were able to complete five cases. This number, obviously, limits our ability to generalize from the cases. We do have the survey data, however, from which generalizations are possible. These data provide measures of the extent to which a given case situation conforms to general industry practice. Case study results that might be at extreme variance with the survey, therefore, could be checked to see if there were unusual circumstances that explained the departure from expected norms. One of the tasks of the research teams was to describe the environment and background of a firm at the time of the design or selection project as completely as possible, to allay possible criticisms that the cases were unique or atypical.

CASE SELECTION

Potential case study opportunities were identified during the survey described in Chapters 1 and 2. Companies surveyed were asked if they might consent to our conducting a case study. The survey respondents had already described at least one instance of the application of advanced manufacturing technology, so we were aware of specific cases that might be studied.

We tried to exercise some discrimination in the types of cases we attempted. We wanted some of the cases to represent the "norm" of corporate human factors policies, that is, situations that tended to fit the central tendency of the responses we received in the survey phase of the project. We wanted other cases, however, in which there was extensive attention to human factors principles in one or more policy area.

Our preliminary screening of possible candidate studies was guided by a set of "go—no go" criteria:

1. The technology involves advanced forms of computer (or programmable controller) applications in manufacturing.
2. We are able to obtain data on what occurred at the design and/or selection stage of the technology.

3. We can have access to people at all levels in the organization who were involved in the project.
4. The project has been completed and the technology is in operation.

We also sought to study cases that had most recently been completed, so our chances of reaching many of the actual participants and of getting access to many of the documents would be greater. Limitations of research funds placed some constraints on where we would be willing to go to make the case studies—we gave preference to opportunities that were near either Boston or Columbus. Having two research teams in different parts of the country was a distinct advantage in giving us broader coverage at reasonable cost.

The companies and divisions in which case studies were made are as follows:

Company	Division	Location
Polaroid	Camera	Norwood, MA
Timken	Faircrest Steel	Canton, OH
United Technologies	Sikorsky Aircraft	Stratford, CT
Westinghouse	Electrical Systems	Lima, OH
Whirlpool	Clyde Division	Clyde, OH

METHODOLOGY

Our first step after selecting a candidate firm for study was to obtain permission from a senior manager or official in the company. In each instance, we gave assurances that (1) during the study, data obtained would be kept confidential, and (2) the company had the right to review the draft of the case study before it was published to determine whether any information of a proprietary or competitively sensitive nature had inadvertently been included.

Each study involved research teams of three, four, or five people. The team visited the installation and got explanations of the important functions of the technology and features of the products being processed by it. The team, together or individually, met with engineering management, design engineers, manufacturing engineers, manufacturing management, manufacturing supervision, representatives from safety and quality staffs, and operators and technicians working with the equipment.

To maximize the amount of information gathered, we assigned "points of view" to each researcher. These points of view included:

1. Machine design and relationship to product
2. Process performance
3. Product aspects, including quality, specifications, rates
4. Machine maintenance, repair, modifications
5. Management and supervision

6. Operators and technicians
7. Financial aspects, justifications, performance
8. Suppliers

These assignments sensitized each individual on the team to observe things that might otherwise have been missed and motivated him or her to record what was observed relative to that perspective. The data gathered came from direct observation, from interviews with people involved with the design, selection, installation, and operation of the technology, and from documents connected with the project. Documents included annual reports, capital authorization requests, postinstallation audits, internal publications, equipment layouts, product drawings, and so forth.

A study usually required three to four days of visits. At the end of each day of visitations, the research team met to debrief and to plan strategy for the subsequent day. Draft write-ups of each aspect of the case were prepared within a week after the plant visits, while memories were fresh. Each writer was instructed to present only factual information gained from the visit. This could include opinions expressed by the people interviewed, but was not to include opinions of the interviewers. Writing the drafts tended to uncover gaps in our information base, and at least one follow-up visit or telephone call was necessary to fill in the missing data.

Part of the schedule of the study included a feedback session with the senior people involved in the project, including the contact person who initially obtained permission for us to conduct the study. During this session we reported what we found (and what we did not find), and our early impressions. This session proved to be a form of accuracy check on our observations, and it also tended to elicit further useful information.

We then prepared a final draft of the case study, which was submitted to the company for review and comment. Where we had inadvertently included data that was proprietary or confidential, we changed the draft in a manner that was acceptable to the company. In each instance where modification was necessary, an attempt was made to preserve as much of the sense of what had originally been written as possible.

The completed case study, containing only data and observations, was then put on file awaiting completion of all the studies, at which time analysis of the studies would be possible. Chapters 5 through 9 are the five case studies. The analysis is presented in Chapter 10.

5

POLAROID CORPORATION VIEWFINDER ASSEMBLY MACHINE

THE COMPANY

With annual sales of $1.8 billion and 13, 000 employees, Polaroid Corporation was, at the time of our case study, the second largest manufacturer of photographic equipment and film in the United States.* Eastman Kodak, its major American competitor, had sales of $11 billion in the imaging sector of its business. The sales of the two companies comprised a very substantial share of the total American photographic market. Foreign manufacturers of cameras and film, particularly 35-mm format, constituted the major competitive threat to the two U.S. leaders.

In its annual reports the company defined its Corporate Profile in this manner:

> Polaroid Corporation designs, manufactures and markets worldwide a variety of products primarily in instant photographic cameras and films, magnetic media, light polarizing filters and lenses, and diversified chemical, optical and commercial products. The principal products of the Company are used in amateur and professional photography, industry, science, medicine and education.

The company had built a reputation on its innovative products. Its business strategy had relied largely on invention, rapid product change, and patent protection, a strategy that was exemplified in recent years in a stream of film improvements and new cameras and in its successful patent infringement suit against Eastman Kodak's instant photography products.

Polaroid Corporation was located primarily in Massachusetts. Its corporate

*Dates of on-site visits: October 24 and 28, November 14, 1988.

headquarters, engineering, marketing, and research staffs were in Cambridge; its film and chemicals operations were in Waltham and New Bedford; and its camera operations were in Norwood. Foreign manufacturing facilities were located in Scotland (Vale of Leven) and the Netherlands (Enschede).

The greatest share of Polaroid's sales were in the amateur camera and film market, but slightly over 40 percent of its sales revenues were from technical and industrial markets.

After a remarkable period of growth from the late fifties to the late seventies, during which sales increased more than 16 percent per year, Polaroid entered a period of little or no sales advance. Only in 1986 and 1987 did the company experience growth of the kind it had known a decade before. Table 5–1 summarizes worldwide sales and profits for the 12 years from 1976 to 1987. Polaroid had been consistently profitable ever since 1949, one year after the first instant camera and film were commercially introduced.

Manufacture of Polaroid's products was highly integrated, particularly with respect to production of any aspect of a product that was technically difficult or that embodied proprietary designs or processes. Thus, Polaroid manufactured its own color negative, batteries, chemical reagents, coated receiving sheet, and other major components of its films. Similarly, it assembled all of its consumer cameras, making key proprietary components, but buying parts that could be made by conventional technology.

As an employer, Polaroid had a reputation for unusually fine human relations. Typical of the company's concern for people is this excerpt from its handbook for employees, printed in July of 1981:

> We have two basic aims here at Polaroid.
>
> One is to make products which are genuinely new and useful to the public—products of the highest quality at reasonable cost. In this way we assure the financial success of the Company, and each of us has the satisfaction of helping to make a creative contribution to society.
>
> The other is to give everyone working for Polaroid personal opportunity within the Company of the full exercise of his talents; to express his opinions, to share in the progress of the Company as far as his capacities permit, to earn enough money so that the need for earning more will not always be the first thing on his mind—opportunity, in short, to make his work here a fully rewarding, important part of his life.

Table 5–1 Polaroid Corporation Sales and Profits, 1975–1987 ($ millions)

	1976	1977	1978	1979	1980	1981	1982	1983	1984	1985	1986	1987
Sales	950	1062	1377	1362	1451	1420	1294	1255	1272	1295	1629	1764
Profits (after-tax)	80	92	118	36	85	31	24	50	26	37	104	116

Source: Company Annual Reports

> These goals can make Polaroid a great Company—great not merely in size, but great in the esteem of all the people for whom it makes new, good things, and great in its fulfillment of the individual ideals of its employees.

This statement of Polaroid's basic philosophy echoed much of what was said 20 years earlier in the company's 1961 annual report to its stockholders:

> From the founding of our Company, it has been our conviction that to be genuinely successful, our Company must not only make worthwhile products, genuinely new and useful to the public, it must also make a worthwhile worklife for every member of the Company.
>
> We believe these two aims are indivisible.
>
> The second objective has shaped the programs that contribute to a challenging and rewarding worklife for all in the Company: our programs for education, training, career guidance, and counselling; presenting opportunities for advancement by posting job openings; our profit-sharing retirement and incentive compensation plans; our experiments, now well-launched, in dual-job rotation.

Polaroid did not have a union. It did, however, have an Employees' Committee, consisting of elected representatives who communicated suggestions and complaints, assisted employees who had grievances, and reviewed proposals for changes in company policies or employee benefits. There were two payrolls, weekly and salaried. Pay was determined by the job grade assigned to each job and by the pay position in the job grade. The job grade was determined by the skill requirements of the job and the nominal pay rate for that grade was set by what would be paid for comparable skill in the industrial community. A person's individual pay position within a job grade, however, could be on any one of seven 5 percent steps, and this was determined by performance, not seniority. Polaroid's policies with regard to time off and a variety of benefits tended to make no distinction between payrolls—all employees were on the same footing.

The divisions of Polaroid of interest to this case study were the Camera Division, located primarily in Norwood, Massachusetts, and the Equipment and Facilities Engineering Division (EFED), with headquarters in Waltham, Massachusetts. These two units were involved in the design, construction, and use of the Viewfinder Assembly Machine, a technologically advanced manufacturing system that automated part of the Spectra camera assembly process.

The Camera Division was responsible for manufacture of Polaroid's consumer camera line, consisting of the Spectra camera for the "high end" ($75 to $150) market, the Impulse camera for the mid-range ($40 to $75) market, and a few versions of the 600-series camera in the "low-end" (under $40) market. The 600-series cameras had been highly popular in the early 1980s, but were now approaching the end of their product cycle. These camera products and their accessories were assembled at plants in Norwood and in the Vale of Leven, Scotland.

Employment in the Camera Division in the United States in 1988 was 850, of whom 300 were salaried, and 550 were direct labor personnel. Employment levels in the early 1980s had been as high as 2200. The division was primarily a manufacturing arm of the company for high-volume production of cameras. Product design was handled by a separate engineering organization in Cambridge while camera sales, marketing, and distribution were located in Cambridge and Needham, Massachusetts.

The Camera Division organization had a Director, to whom four Plant Managers (Optics Manufacturing, Electronics Assembly, Camera Subassembly, Camera Final Assembly) reported. Staff managers of Human Resources and Finance also reported to the Director. The hierarchy of line management in each plant consisted of a Production Manager, several General Supervisors, and, at the lowest level, Supervisors. Each of the plant managers had staff organizations for Materials Management (purchasing, production planning, inventory control), Manufacturing Engineering, Quality Assurance, Finance, and Human Resources. Many of these staff organizations had duplicate reporting relationships to people either at division or corporate levels. Each Quality Assurance manager, for example, had a "dotted line" relationship to the corporate Quality Assurance office. Each Human Resources manager reported to the divisional Human Resources office.

The Equipment and Facilities Engineering Division (EFED) was independent of the Camera Division. For many years it had been responsible for the design and selection of all of Polaroid's manufacturing equipment. Most of the machines for proprietary processes were designed and built in-house. The division had a large staff of experienced engineers and extensive machinery construction shops. A large portion of the equipment designed by EFED was completely fabricated, assembled, and installed by groups within the EFED. In past years, this in-house capability had permitted Polaroid to maintain close control of information about new product start-ups and proprietary processes.

The EFED organizational structure was horizontally oriented, with eight functional activities reporting to the Division Director. These eight activities were:

Assembly systems and process engineering
Development
Facilities
Electrical controls and instrumentation
Construction and administration
Chemical engineering
Finance
Capital planning

In the case we studied, both design and construction of the machine, which incorporated advanced forms of manufacturing technology, were accomplished within the EFED. Responsibility for the Viewfinder Assembly Machine was assigned largely to the Assembly Systems and Process Engineering Department, but other engineering groups (some external to EFED) also participated in the design.

The population of the EFED at the time of the case study was 650, of whom roughly 300 were engineering professionals. The EFED headcount had been approximately the same two years earlier, at the time of the design and installation of the Viewfinder Assembly Machine.

THE PRODUCT: VIEWFINDER FOR THE SPECTRA CAMERA

The product that motivated the design, construction, and installation of the viewfinder assembly machine was the Spectra camera, introduced in 1986 to capture market share in the $100 to $150 end of Polaroid's amateur camera line. The camera was only part of what Polaroid termed the Spectra System, which consisted of a new film format with new photographic chemistry, the camera, and supporting products and services. Polaroid's 1985 Annual Report, published early in 1986, had this description:

> The Spectra System comprises a computerized camera that focuses and controls exposure automatically; an instant color film that uses two imaging chemistries; a host of accessories, including a remote control device; and Polaroid's laser print service which provides computer enhanced, laser-produced prints and enlargements of Spectra System photographs.
>
> The camera uses advanced computer-like circuitry to make more than 30 complex focusing and exposure decisions within 50 thousandths of a second. New optics and new electronics in the shutter and exposure system extend Spectra System's capability to achieve high quality results over a wide range of photographic situations.
>
> Spectra film combines two separate imaging systems which work in harmony to eliminate unwanted photographic interaction within the film's light sensitive element. This pairing of film chemistries produces unprecedented color fidelity, brightness, and range in an instant film. The final print is delivered in a rectangular format with an image area larger than 600 series film.

An artist's rendition of the Spectra camera, open and ready to take pictures, is shown in Figure 5–1. The camera could be folded down in clamshell fashion to protect the front elements of the camera when it was not in use. Polaroid's announcement of the camera introduction claimed that during its design a great amount of attention was given to making the camera's "feel" and operation ergonomically satisfying:

> Every attempt has been made to make the camera fit the diversity of the human form. Every facet of it—wedge shape, color, weight, texture, control panel layout, shutter button position, instructional symbols—results from an extensive analysis of human measurement data and from repeated research with consumer focus groups . . . From the inception of instant photography, Polaroid has taken a systems approach in which the design for the film and camera complement each other. The Spectra System is no exception. It is the new generation of an instant photographic system that is designed to meet the specific needs of the consumer photographic market.

Figure 5–1 Polaroid Spectra camera

In addition to having a new exposure system, shutter, electronics, and optical system, the Spectra camera was equipped with a novel viewfinder. The viewfinder not only gave the photographer a clear image of the picture to be taken, but it also provided light-emitting diode displays of camera-to-subject distance, whether the strobe light was needed, and whether conditions were right for a good picture.

The Spectra viewfinder design was representative of the lengths to which the camera designers had gone to produce a combination of technical achievement and aesthetic expression. The viewfinder's optical system, for example, consisted of six lenses requiring an optical path of more than 6 inches. This sophisticated system was fitted into a camera body where the front-to-back viewfinder length could be only 4½ inches. The resulting design (Figure 5–2) required light from a scene to perform gymnastics off four mirrors before it emerged at the eyepiece. Polaroid's description of the viewfinder was as follows:

> The Spectra System's 21-piece viewfinder is a model of functional design, optical precision, and constructional simplicity. This new viewfinder has exceptional brightness and sense of scale. The viewfinder simplifies viewing and framing the image and provides a sharply defined, black subject mask. LED readouts in the viewfinder inform the photographer about exposure conditions and camera status and, in the viewfinder's preview mode, indicate when conditions are satisfactory for photography . . .

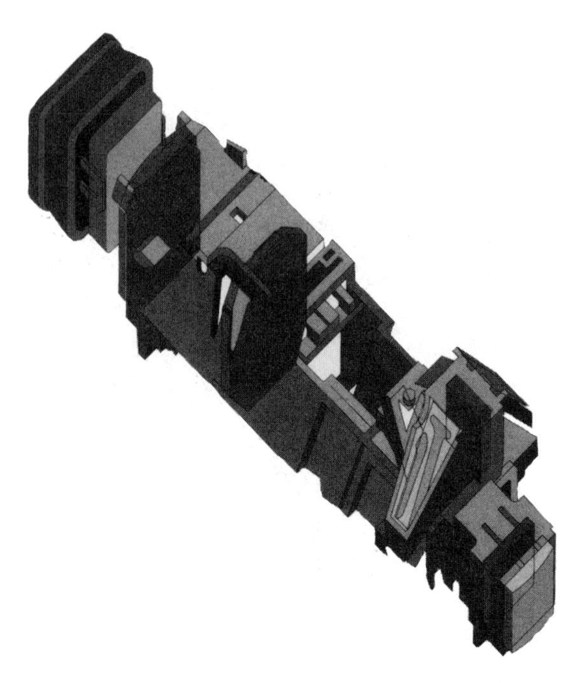

Figure 5–2 Spectra viewfinder

> The viewfinder is itself a sophisticated optical system: six precision plastic lenses and one prism with five aspheric surfaces, four mirrors to erect the image for right-side-up viewing, and a reflecting prism for viewing the data displays. Its front window has been placed close to the camera's taking lens to minimize parallax when shooting subjects at close range.

All elements of the viewfinder were assembled into a boxlike molded plastic housing. Component parts, primarily molded plastic pieces, were snapped into position. There were no fastening screws, rivets, or similar devices. One adjusting screw was the only conventional piece of hardware. An exploded view diagram of the viewfinder parts is shown in Figure 5–3.

The complexity of the viewfinder design, with its many optical surfaces that could collect dirt or fingerprints during assembly, virtually mandated that the assembly process be automated. Added impetus for automation came from estimates that it would take 140 people to assemble the viewfinder manually in the quantities required. Because Polaroid had been trying to maintain continuous employment for its employees, it did not want to enlarge its work force for this product, when later fluctuations in demand might cause layoffs. So three considerations—complexity, cleanliness, and employment policies—drove the process designers to seek approval for an automated approach to viewfinder assembly for the first time in Polaroid's history.

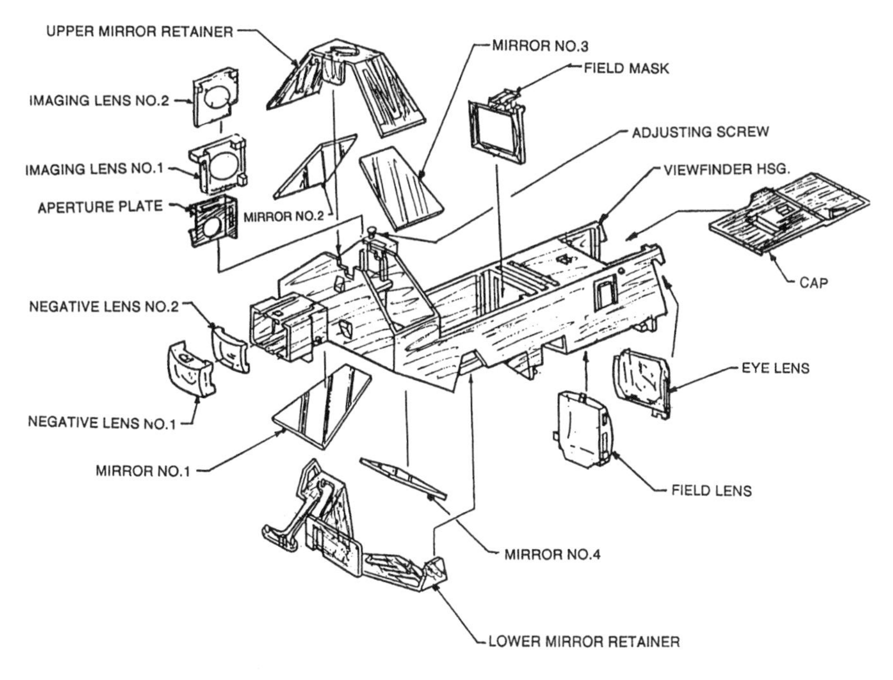

Figure 5–3 Viewfinder assembly diagram

THE PROCESS: VIEWFINDER ASSEMBLY MACHINE

The viewfinder assembly machine performed all the necessary steps to assemble, test, adjust, and inspect the Spectra viewfinder. Over the three years of design and installation, the machine evolved into a one-of-a-kind complex of advanced technologies that included robots, programmable controllers, ultrasonic equipment, servomotors, indexing parts feeders, sensors, and automatic inspection stations. The approximate length of the assembly line was 50 feet. It was served by a crew of seven on each shift.

The assembly line consisted of 11 machine sections, or modules, linked by short transfer mechanisms. A flowchart of the process and a schematic view of the line are given in Figures 5–4 and 5–5.

The molded plastic viewfinder housing was the major structural part to which all other parts were assembled. Along the length of the line a series of "nests," or static holding fixtures positioned the housing for whatever assembly or test operation was to be performed at that stage. At each module, housings were moved from nest to nest by a "walking-beam" mechanism, a three-foot, cam-actuated, horizontal member that picked up the housings, carried them the distance between nests, deposited them into their next fixtures, and returned to its original position for the next lift-carry-drop cycle. Six short, three-position conveyors carried the housings from one module to the next.

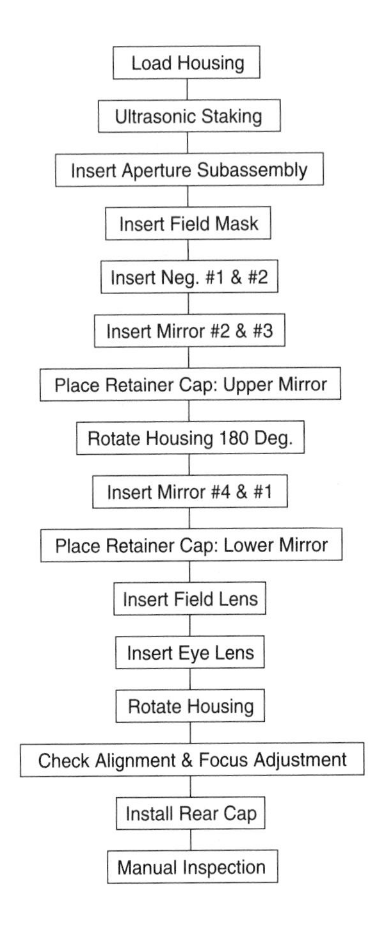

Figure 5–4 Viewfinder assembly flowchart

Housings were manually inserted in proper orientation in the first of the 134 nests on the machine. From that point on, all parts were added to the housing by machine. When the machine was operating properly, there was no human intervention in the assembly process. People loaded parts into the machine, monitored its performance, and cleared jams. As was intended by the machine designers, no human hands touched any of the optical parts during assembly.

When an assembly arrived at a particular workstation, either a robot or other mechanical device performed the work. There were a total of eight robots in the line, all Unimation PUMAs, models 200 and 260. The robots had six-axis rectilinear configuration. The tools and end-effectors were custom engineered to do specific tasks. Most of them were vacuum grippers, using suction to pick up lenses and mirrors. These small robots were capable of about six pounds of force. Each robot joint had its own feedback control system. Eight microprocessor-based controllers were programmed to coordinate the motions of the robots. Each of the robot controllers and the corresponding section of walking beam were, in turn, coordinated by a Modicon programmable controller.

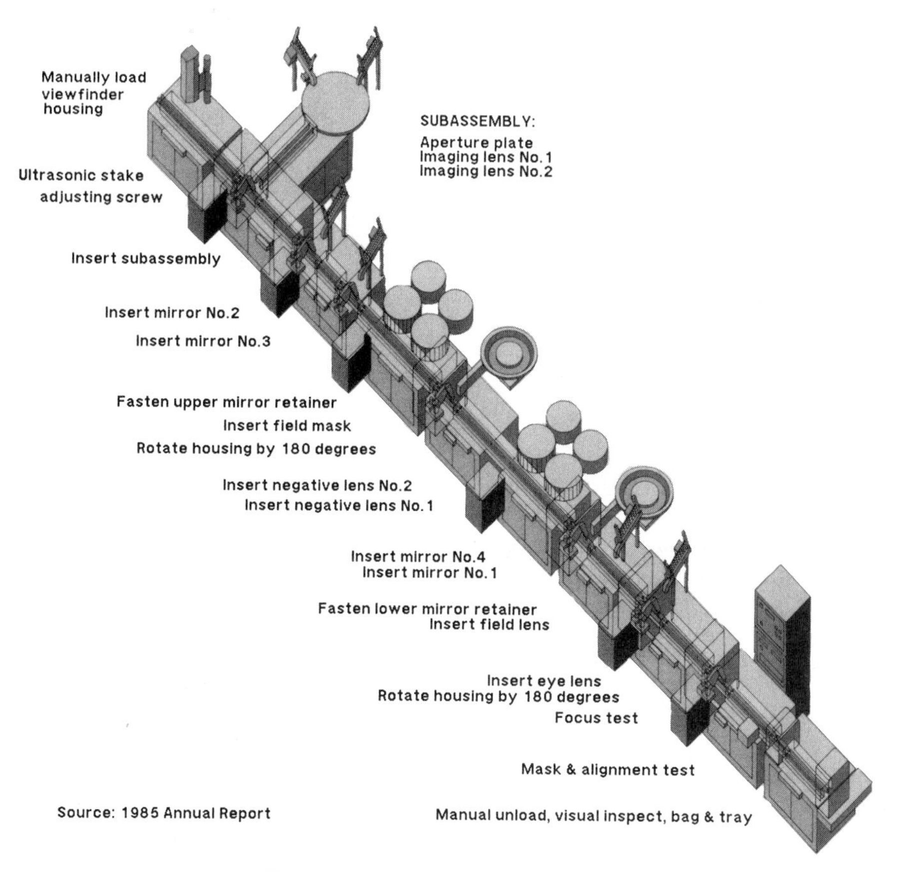

Figure 5–5 Viewfinder assembly machine

Parts feeding was done from the back side (right side of the diagram in Figure 5–5) of the machine. Nonoptical plastic parts were oriented and presented to the robot pickup devices by vibratory bowl feeders. Oscillatory motions of a bowl in which loose parts were placed caused the parts to migrate to the periphery of the bowl and to climb a spiral ramp around the inside wall. Only correctly oriented parts could make it to the top, where they would be picked up by a robot or other transfer device. Parts improperly oriented fell back into the bowl to start the climb anew.

Optical parts—lenses and mirrors—were loaded into special cartridges at the location where they were fabricated before being moved to the assembly area. Lenses were placed in clear acrylic tubes similar to the DIP (dual inline package) tubes used in integrated circuit insertion machines in the manufacture of printed circuit boards. Mirrors were stacked in similar fashion in metal frames. Use of the cartridges maintained parts orientations and ensured that the optical surfaces were kept clean. Tubes of parts were loaded manually into magazines at the top of the machine. Each tube in succession would be emptied and would fall into a bin.

Three operators handled all parts loading. One continuously fed housings into the first stage of the machine, while the other two loaders made certain that adequate stocks of all other parts were on the machine. For most parts, replenishment was required about twice per shift.

The six short transfer conveyors that were located between certain of the modules gave some degree of buffering in instances where a part might jam in one station and require that that module be stopped while the jam was cleared. The three parts in the conveyor could sustain subsequent operations for a short time before those operations would also have to shut down. Sensors at various locations on the machine prevented the machine from trying to cycle when housings or other parts were missing. When there were no interruptions for parts jams or other reasons, it would take a housing about 10.6 minutes to emerge from the machine as a completed viewfinder assembly.

Assembly Operations

Short descriptions of the sixteen operations performed along the assembly machine follow. These provide a somewhat clearer idea of both the complexity and the interdependency of the assembly system.

Load viewfinder housing. The housing was placed manually in a nest from which the walking beam carried it forward to subsequent operations. The procedure did not involve great dexterity on the part of the operator, but did require continuous repetitive motion. The complexity of the housing design and the need to avoid any downstream jamming called for careful seating of the parts. Attempts had been made by the machinery engineers to design an automated feeding system, but lack of both space and an effective solution led them to resort to the manual method. The operator had a control switch to stop the line if necessary and could also carry out the steps needed to restart it.

Insert and stake adjusting screw. Alignment and focusing in the viewfinder were accomplished by tilting a mirror with the small adjusting screw introduced at this station. Loose screws were loaded into a vibratory bowl feeder, and properly oriented individual screws were picked up and placed into a starting hole in the housing. An ultrasonic staker then forced the screw into the hole, creating a threaded hole for later adjustment of the screw position. The person loading housings at station 1 kept a casual watch for jams in the bowl feeder and used a scriber to disengage any interlocked screws. Jams at this point were quite infrequent. Sensors were provided at the exit of the station to check whether the operation was complete. Sensor accuracy was important, because a defect missed at this point would pass through the entire length of the assembly line to final inspection.

Insert aperture subassembly. A subassembly operation and an assembly operation were both performed at this stage. Imaging lenses #1 and #2 were combined with the stamped metal aperture plate to form a subassembly. The subassembly was then inserted into the housing. A rotary indexing machine put together the subassembly. Lenses were fed to the rotary table from plastic tubes; the aperture plates were sorted and presented by a vibratory bowl feeder. The completed subassembly was picked up from the rotary table by the first robot on the line and inserted into the viewfinder housing. At the end of the station a probe searched for the presence and height of the aperture plate assembly to confirm proper positioning.

Insert field mask. The field mask framed a view exactly as it was to be in the picture itself. The mask was installed by the second PUMA robot on the line. The robot took the part from a gravity feed track and, with a vertical downward motion, inserted the mask into the housing. Slight pressure exerted by an air cylinder rotated the part into its final position. Because of the snap-action assembly, control of the snapping force was an important design consideration.

Insert negative lenses #1 and #2. A third robot inserted negative lenses #1 and #2. Its end-effector had two fingers with suction cups. The longer finger picked up the smaller lens (lens #2) from the feeder and inserted it first. A swiveling action brought the shorter finger in contact with the bigger lens, which it picked up and placed in the housing. Slits in the housing provided elasticity for the snap action. A sensor at the end of the station checked for the presence of one of the two lenses.

Install mirrors #2 and #3. At module 4, a robot placed both mirrors #2 and #3 in the housing. The small glass mirrors, prepared in another area of the plant, arrived at the assembly area stacked in teflon-coated metal tubes. The surface of each mirror had been coated with polypropylene to protect it prior to assembly. The mirror tubes were loaded into a rotary parts-feeding table that supplied mirrors to a second indexing table where the polypropylene coating was stripped off by heated air and suction. Fumes created by this operation were filtered before the air was discharged into the room. Proper control of the hot-air temperature was necessary because overheating of the coating layer caused melted plastic to adhere to the mirror surface, and underheating left the layer intact.

Defective mirrors placed in housings at this station were detected only at the focus adjustment station in module 9, so there was a possibility that the entire line from this point could become loaded with defective assemblies. The lead technician was given the responsibility of regulating the temperature. Two rotary tables presented the mirrors to a single robot that inserted them sequentially. After each mirror was inserted, an infrared sensor checked for part presence. A sensor on the robot arm checked for the presence of a housing in the fixture, and, if the cavity were found vacant, the robot arm would not release the mirror, and the rotary tables did not index.

Fasten upper mirror retainer. The upper retainer cap snapped into position on the housing and held mirrors #1 and #2 in place by spring force. PUMA robot 4 picked up the mirror cover from the end of a bowl feeder and snapped the part in place. Jams at the bowl feeder would prevent the robot from picking up a piece and would cause the module ahead of the station to stop. Parts loaders kept watch for areas of probable jams.

Rotate housing 180 degrees. Mirrors #1 and #4 had to be inserted into the underside of the housing, so the operation at this station was to turn the housing 180 degrees. This step made it possible for the robot at the next station to insert the mirrors from above the housing.

Install mirrors #1 and #4. After the housing had been rotated, mirrors #1 and #4 were inserted by the robot in the same manner as described in operation 6, installing mirrors #2 and #3. The only difference was that this station lacked a sensor at the end of the robot end-effector. Whenever there was an empty nest because a housing was missing, the robot would proceed to deposit mirrors in the empty cavity. The design engineers were aware that the station lacked the capability of sensing when nests were empty and were in the process of remedying the problem.

Fasten lower mirror retainer. At this station a robot attached the lower mirror retainer, which captured mirrors #1 and #4. The retainer cap had an intricate shape that required forces from two directions to snap it into place.

Insert field lens. Instead of a robot, a simple pick-and-place machine inserted the lens into slots in the housing and pressed it in for a snap fit.

Insert eye lens. The eye lens had to be pressed into the housing from an awkward angle for snap fitting. Full six-axis articulation of the PUMA robot was utilized to snap it into place. At an earlier stage in the evolution of the assembly machine, this operation had been a manual job that had given operators hand and wrist problems. Lenses at this and the preceding field lens-insertion station were fed from tubes of parts. One of the two operators responsible for parts loading replenished the stacks of tubes when necessary.

Rotate housing 180 degrees. A mechanism similar to that in operation 8, the first housing rotation, restored the housing to its upright position.

Test focus and adjust alignment. Two devices performed automatic final inspection and adjustment of the viewfinder. The first device checked the alignment of the optical path, trimming it by means of the adjusting screw and by shifting the field mask laterally across the optical axis. The second checked the optical system for focus. Failure of focus meant a serious defect, usually a missing or faulty part. When a defective viewfinder was found, its identity was signaled to the computer, which kept track of it from that point to the end of the machine. There the faulty assembly was diverted to a repair/rework accumulator. The station was connected to a small computer that accumulated production information on output, reject rates, machine downtime, and so forth. None of the operators had access to the software for that computer.

Install rear cap. The rear cap was snapped into place on the housing by a simple mechanism (not a robot) that was supplied parts by a vibratory bowl feeder.

Divert faulty viewfinders, manually inspect, bag, and tray good viewfinders. Based on information generated at the focus test station, a solenoid-actuated gate at the end of the machine would deflect all faulty viewfinders to an accumulator for later manual repair or rework. At the time of the first visit of the study team to the machine area, good viewfinders were passed on to three operators who visually checked the assemblies, cleaned them with a jet of air, bagged them, and placed them in a tray for transport to camera final assembly. Rejected assemblies were sorted for rework or scrap. When a sufficient number of assemblies required rework, an entire shift was devoted to correcting them. In the month following the visit, the three inspection positions were eliminated because of the consistently high quality of the assemblies. This step reduced the total operator-technician crew to seven persons.

THE DESIGN PROCESS

The Spectra camera design featured both performance and style. Market research had identified those features that customers wanted with respect to the shape of both the camera and the finished photograph. The designers in Product Engineering took into account customer preferences for shutter release placement, viewing, focusing, and overall camera shape. They gave careful consideration to ergonomic design of the camera, so that the product fit the hands and face of the photographer. These features added to product appeal, but they created severely limiting constraints for the internal mechanisms of the camera.

A major camera feature was a sonar range finder coupled to an automatically adjusting lens system that could focus on objects at any distance from one meter to infinity. To match the capabilities of that system, the designers chose a telescopic viewfinder design that would enable a photographer to see the scene being photographed clearly and in focus over the whole distance range. As was mentioned in the section on camera design, fitting the viewfinder into the camera body tested the designers' ingenuity. Polaroid did have a computer-aided design (CAD) system. Without it, the camera could not have been completed in the time allowed, according to one of the senior design managers.

When confronted with early concepts of the viewfinder, the Optics Manufacturing organization in the Camera Division was told merely that tolerances for the system had to be "very tight." Although "very tight" remained undefined for a long time, it was clear to the manufacturing people that all components of the viewfinder subassembly would have to be of the highest dimensional precision and optical quality. There could be no scratches or fingerprints on either lenses or mirrors. Precise placement of parts and very clean assembly practices would be necessary to meet the product design requirements.

In 1983 Optics Manufacturing was asked to provide early cost estimates for fabrication of the viewfinder subassembly. Working from very sketchy information, Camera Division industrial engineers prepared figures on the basis of conventional manual assembly techniques. The resulting numbers were considered too high relative to the projected price of the camera. Camera production costs, according to Camera Manufacturing management, were "an overriding factor" in all considerations. The estimated viewfinder costs were "out of line" with all other costs. Manufacturing had to find ways to cut the costs.

Camera Manufacturing requested that EFED be given the responsibility for developing a process that met the cost and quality goals. What emerged was a concept of a totally automated assembly line in which no part was to be handled by humans, once it had been loaded into the machine. The final design, complete with automated transfer and robotic assembly, has already been described. This design, however, did not arise full-blown out of EFED. It emerged in a series of evolutionary steps, prodded by top management's insistence that the latest in flexible assembly technology be employed.

EFED's first design approach concept was to go to hard-wired automation: dedicated pick-and-place devices, rotary parts-feeding tables, and a walking-beam mechanism to transfer the assembly down the line. Each assembly step was to occur at a workstation module that was linked by a transfer mechanism to the preceding and succeeding stations. All part manipulation and assembly was to be accomplished by inflexible automation. This was a low-cost solution to automated assembly. It was also a technology that was familiar to engineering groups in EFED.

When the EFED presented their plan for design and construction of the automated viewfinder assembly machine to top management for capital funds authorization, it was rejected. The president of the company insisted that the design incorporate flexible automation, using robots. The directive was to develop "a way of bringing the company into the twentieth century."

The revised design had a combination of pick-and-place devices and robots, and this was approved. The financial justification for the system compared the expected costs of manual assembly (134 operators) with those of the proposed automation. Further justification was based on the ability of the proposed system to meet quality and quantity requirements, and the increased flexibility that would be afforded by a robotic installation. These decisions were made in late 1983.

Equipment design efforts escalated after this decision. There quickly evolved close, but frequently strained, working relationships among the groups responsible for implementing the design concept. The groups included EFED, Product Engineering (responsible for camera design), Optical Engineering (part of Product Engineering, responsible for camera optics), and Camera Manufacturing (represented by managers and engineers from Optics Manufacturing and by divisional industrial engineering). When representatives from these groups were asked what prompted the close relationships, the consensus response was "survival."

In this instance, "survival" meant personal professional survival at Polaroid. The Spectra camera project was such a major undertaking for the company that each part of the program *had* to succeed. For EFED, this was the first application of robots in automated assembly in the company, so they had to "do it right." Meeting the product introduction date of mid-1986 was crucial, so the process designers had both performance and schedule constraints. One further complication: There was competition for scarce engineering resources in EFED because the Spectra program had a number of other high-priority projects.

The prospect of designing a flexible assembly system for the viewfinder was met with mixed emotions by people in EFED. The division had been looking for an opportunity to design an automated assembly process that could survive model changes. They had even selected a camera subassembly (the motorized film transport drive and roller system), and were considering equipment alternatives that might be used for flexible automation. When the viewfinder project was activated, work on the camera drive assembly machine was set aside. The new project may have been an opportunity, but few engineers in EFED had had any experience with robots. "The older engineers feared them," was one comment. With top management insisting on a robotic application, however, the engineers set about to develop a workable system.

Design Principles

When asked what were the basic design principles that guided the viewfinder assembly machine project, the senior designers involved came up with this list:

1. Keep it simple.
2. Keep it safe for operators.
3. Design for reliable performance.
4. Design for ease of maintenance.
5. Use commercially available mechanisms (feeders, robots, stakers, etc.) whenever possible, and standardize on a few suppliers.
6. Analyze design for probable fault.
7. Maintain close relationships with the product designers.
8. Have the machine tell the operator why the machine malfunctions.
9. Provide automatic counting of product for the management information system.
10. Keep decision making away from the operator—the operator should run the machine but make no adjustments.
11. Performance goals: 95 percent yield of good product, 95 percent uptime.

Two early design decisions also guided events for the three years from 1984 through 1986. One decision was to design the assembly system as a set of modules that could be individually assembled and debugged, rather than to design a monolithic machine. The second decision was to install and operate the system in phases that permitted a gradual shift from manual assembly to complete flexible automation. At intermediate stages, "hard-wired" automation would be used until it could be replaced by robotic technology.

Design Chronology

A rough sequence of events in the design, construction, installation, and operation of the viewfinder assembly machine is as follows:

1983 (early)	—Spectra camera design, early machine concepts.
1983 (late)	—Machine project approved.
1984 (early)	—Machine design begins.
1984 (late)	—Construction of modules begins.
1985 (July)	—Installation of modules making up the walking-beam line.
1985 (October)	—Start of manual assembly of viewfinders on the walking-beam line, and at separate workstations; up to 40 operators involved, on a single shift.
1985 (late)	—Installation of pick-and-place machines.
1986 (early)	—"Hard-wired" automation at four workstations; three-shift operation, 11 operators per shift.
1986 (October)	—Installation of eight robots, conversion of line to flexible automation; four operators, three inspectors, three repair/adjust operators, one lead technician per shift.
1987–1988	—Modifications to line while operating; improvements in performance by end of 1988 permit reduction of crew size to total of seven per shift.

This sequence did permit the project team to meet the deadline for camera introduction, but full flexible automation was not accomplished until well after the Spectra system had been introduced to the public. The machine performance goals of 95 percent yield and 95 percent uptime had not been consistently met at the time of the case study in late 1988.

Problems Encountered

The creation and successful operation of the viewfinder assembly line was the result of teamwork between design and manufacturing groups at Polaroid. Because automation of the line was done in a series of stages, transfer of full responsibility for its operation from engineering to manufacturing took more than two years to complete. Problems arose at all stages, from design to operation. During the design phase, the team had challenging constraints that had to be satisfied and shortcomings that had to be surmounted:

- The machine had to accommodate the complex shape and structure of the Spectra viewfinder. Because of "real estate" problems in the camera, the viewfinder was given an intricate shape with periscopic features. Tight tolerances imposed on the components of the viewfinder became translated into similarly tight tolerances in the assembly machine.
- Capacity requirements dictated that the machine be able to run reliably for three shifts a day, requiring speedy maintenance and accurate product fault detection.
- Cost constraints mandated high yields and low labor cost.
- Personnel policy specified that additions to the Polaroid payroll be kept to a minimum. Automation was indicated.
- The designers had limited experience with robots, and no experience with combining a number of robots on the same machine. This situation was further complicated by the fact that the robots had to be incorporated into the machine after it had already been installed in manufacturing and was producing assemblies.
- Because a completely new camera and film system were to be introduced, many new machine projects were underway at Polaroid simultaneously. Machine design talent was a scarce resource.
- The machine completion date had to be met.

The challenges posed by these constraints were all mastered during the design and construction phase. Many problems were encountered, however, after the machine had been installed and debugged. Some of these problems remained at the time of the study:

- The design decision to eliminate or minimize operator interaction with the machine essentially locked the operator out of all functions other than parts loading. Machine elements, including those that frequently experienced parts jams, were not easily accessible. Clearing jams required either that the affected module of the machine be stopped or that the operator reach into the machine in violation of safety rules.
- Because each robot was independently controlled and not linked into a central data net, lead technicians had difficulty monitoring performance of the eight robots. The study team was told that EFED design policy for future equipment of this type would be to unify monitoring data at a central station.
- The assembly machine had a high frequency of parts jams. This was particularly evident during start-up at the beginning of a shift. Because the machine was operating only one shift a day at the time of the study, start-up was a daily affair. As a consequence, sample data showed the machine was running only about 55 percent to 70 percent of the shift.

- As was evidenced by the frequency of jams, certain parts-feeding mechanisms were having problems. These included feeders for aperture plates, rear caps, and several of the lenses. The mechanism assembling the aperture plate subassembly (three parts) was especially troublesome, requiring almost constant monitoring. At the lens-insertion stations, plastic tubes containing properly oriented lenses were distorting in the stack, holding up the free flow of lenses into the machine. Some vibratory bowl feeders had to be redesigned to reduce the frequency of jams and to present parts in correct orientation.
- The two mirror-insertion stations were the major bottleneck in the assembly line. At least one of the two lead technicians gave virtually undivided attention to these stations, adjusting the temperature of the hot air used to remove the polypropylene coating from the mirror surface, inspecting mirrors to see that the coating had been completely removed, cleaning out dust generated by the process, and checking for mirror damage.
- Additional sensors were needed. A robot that lacked a means of determining whether a viewfinder housing was in a given cavity on the walking beam would release mirrors into empty cavities. The operator had difficulty retrieving these mirrors while the machine was in operation.
- Failure of sensing equipment was not easily or automatically detected. Malfunctioning of one of the sensors could cause the entire line to fill up with defective assemblies that would be rejected at the final inspection stage. Lead technicians had to check the sensors periodically for proper functioning.
- Loading of tubes of parts into the machine stations was made difficult by the height of the parts-holding racks above the machine. It was awkward to make the reach, even though the weight of the parts and tubes was modest.

Many other problems were encountered and solved, and production efficiency gradually improved. The machine continued to produce viewfinders in time and in sufficient quantities to meet camera production requirements for both the Spectra and Impulse cameras. Thus, despite the difficulties, the process design team satisfied the basic goals set for it.

Performance of Viewfinder Assembly Machine

Design of the viewfinder machine continued to evolve for two years after initial installation. Because of this, performance also showed evolutionary progress. The most important performance criterion satisfied by the system was that it continually supplied good viewfinder subassemblies for the Spectra camera on time and in sufficient quantities. Camera production was never held up for lack of viewfinders. The study team obtained information on various aspects of the machine's performance. These are discussed in the sections that follow.

Capacity. The theoretical capacity of the assembly line was nearly 6000 pieces per 440-minute shift. This theoretical figure does not make allowance for losses incurred because of machine jams or quality problems. A computer connected to the machine control system printed out data on key performance measures at the end of each production shift. Actual production varied for each shift, but the study team was able to make a rough estimate of actual system output from a group of randomly selected printouts provided by the company. The sample data indicated an average output of 3000 assemblies per shift with a range from 2300 to 4100 units.

The primary cause for loss of capacity appeared to be time lost in clearing jams. Efforts were being made to eliminate the factors causing jams by redesigning some modules and by increasing operator intervention.

Cycle time and yields. Data for the period from August 1987 to November 1988 showed steady increases in machine performance in terms of increased efficiency, overall yield, focus yield, and cosmetic yield. Cycle time during this period decreased toward the design goal. As a consequence, machine throughput time (the time required for a part entering the first station of the process to emerge at the other end as a completed subassembly) also decreased. At the time of the study, throughput time was approximately 11 minutes. The design goal was closer to 10 minutes.

Downtime and system losses. Downtime statistics varied from one day to the next. Sample statistics indicated that the machine was in full operation (all modules up) about 250 minutes out of a 440-minute shift. Maximum uptime in the sample was 341 minutes in one shift.

Downtime was directly related to the frequency of stops. Only one shift was being run at the time of the study, so there was a greater than normal amount of setup time and frequency of jams during start-up. If the machine had been operating three shifts per day, the setup time between shifts would have been virtually eliminated.

Modules 5 and 7, where mirrors were inserted, had the greatest amount of downtime. The number of stops per shift was as high as 100 at these stations because of jamming. Robot recalibration was also occasionally needed.

Because of these problems, operators had to monitor the process closely and remove defective pieces well in advance of possible jams. They also retrieved dislocated partial assemblies in order to reintroduce them at the beginning of a particular operation. This enhanced the yield and helped the shift to reach its production target.

System flexibility. After the Spectra camera had been introduced, Polaroid's camera designers proceeded to design the Impulse, the second camera in the new line. Now the designers faced a new constraint: the viewfinder assembly machine. The designers had to be convinced by Camera Manufacturing that they should design the viewfinder for the new camera so it could be assembled on the machine. This approach differed from the Spectra project where process design had to accommodate product design. At the time of the study, the line remained dedicated to these two virtually identical products. According to the Engineering Manager of the project:

> When the Impulse camera was being designed, two people working on the product design came down to see us, to check if we could shorten the viewfinder. After they looked at the machine, the design of the viewfinder remained the same.

Commenting on the visit of the designers to the assembly line, the Director of EFED, who had formerly been in charge of Spectra and Impulse camera design, said:

> The plan (for Impulse) was to get a superb viewfinder at low cost. The Impulse also was a heavily style-driven camera. There was pressure from marketing to make it shorter, so that visit [to see the machine] was really a condescension on their part. [The difficulty of modifying the process] was convincing, and we decided against shortening the viewfinder.

The product life of the Spectra camera had been estimated to be eight years, and the cost justification for the assembly machine had been based on this estimate. When asked by the study team how much of the assembly machine could be reused for other assembly tasks, the Engineering Manager stated that the figure might be 60 to 70 percent. He cautioned, however, that rapid changes in robotics technology could quickly make the entire machine obsolete. In making the move toward robotics, EFED project managers had the choice of buying higher priced, more flexible robot technology or of selecting simpler models. Cost considerations and EFED's limited experience in robotics led them to adopt the relatively simple PUMA technology.

Maintenance. There were two levels of maintenance for the viewfinder machine. The first level, handled by lead technicians, involved clearing of jams, resetting of controllers and rotary tables, and reindexing robot arms. In addition to maintaining general cleanliness, the technicians made sure that the walking-beam area was free of contaminants. Parts loaders did not participate in any maintenance, but watched for jams in bowl feeders and other dispensing equipment. Once a week, a drill was carried out by the production crew to clean the line, lubricate moving parts, and check for evidence of wear. None of the crew disassembled any part of the machine or had access to the control software.

A second level of maintenance, which involved disengaging locked air cylinders and other critical repair work, was carried out by trained mechanics. Mechanics also handled all electrical maintenance. If there was a computer problem, the engineer responsible was usually summoned.

Safety and health. From a safety and health standpoint, machine performance included these features:

- Noise level was less than 85 dba.
- The maximum weight handled by a person was 50 pounds.
- The machine could be oiled without removing safeguards.
- The machine layout allowed room for safe maintenance.

The presence of these features was formally acknowledged by Camera Division management when the machine was transferred from EFED to them.

Project interactions. During the period of the most intensive design activity, from early 1984 to early 1985, there were 13 to 14 people from EFED involved in one way or another. As the project progressed during the implementation stage, from early 1986 to product introduction in mid-1986, the effort was shared about equally by EFED and Camera Manufacturing. Finally, by mid-1988, the installation was turned over formally to Camera Manufacturing. By this time, the design engineers were eager to be relieved of their responsibilities so they could go on to new projects.

There were many early frustrations with the delivery of equipment and with the allocation of resources to the project. Project costs, including those incurred by EFED, were being charged to the Camera Division, so the manufacturing participants were concerned whether time and efforts were being applied conscientiously. There were frequent changes to the designs prior to their release to the construction and installation groups. Many of these change requests came from manufacturing, to which EFED had to respond.

The key to resolving design and implementation issues came with the institution of weekly meetings between Optics Manufacturing and EFED. These were described as clear, open, and frequently "at a high decibel level." The tenor of these meetings apparently reflected the stress of trying to meet a fixed implementation date with an unfamiliar technology. At the time of the case study, however, there appeared to be a considerable pride in accomplishment and camaraderie among those who had been involved in this project. One individual remarked that this project had been the highlight of his career.

Responsibility within the Camera Division for assembling the viewfinder shifted during the design of the machine. Because the optics were to be high-precision acrylic lenses, their manufacture was to be done in the Optics Manufacturing plant, which had state-of-the-art technology for producing molded acrylic lenses. Assembly of the viewfinder was originally to have been done in the Camera subassembly plant, located at the same Norwood site but in a separate building. As the design progressed, however, it became clear that the problems of optical quality and precision would be better handled within the Optics Manufacturing plant. Optics Manufacturing people, moreover, had demonstrated that they were helpful in reconciling differences that arose between Optical Engineering personnel in Product Engineering and the machine designers in EFED.

There was no involvement of production operators or technicians during the early stages of design and development of the machine. As one engineer put it, "There was no viewfinder assembly crew before the machines were installed."

When the basic machine modules with the walking beam had been installed and operators were producing viewfinders manually, there was still very little operator participation in identifying design problems or suggesting changes. When the designers were working on incorporating robots in the line, operators were invited to visit a laboratory installation of robots at a non-Polaroid site. The intent of this visit was to reduce operator apprehension about the soon-to-be-installed robots.

The lack of participation of production operators was explained by team members from both EFED and Camera Division as a matter of inexperience on the part of the operators. It was widely believed that the operators would have nothing to offer in the way of suggestions about the equipment or workstations.

The Employees' Committee (elected employee representatives from all Polaroid sites) was given information about the project, but there was no active participation by this group. The Human Resources organization was also not involved in the design or implementation effort. Safety features of the machine, however, received considerable attention. Polaroid's extensive policies regarding safety and their application to the viewfinder assembly machine are described in a separate section below.

The installation of the walking-beam transfer line and the individual assembly modules began in early 1985. The line, which became known as "the beam," was assembled in place on the factory floor. As the product introduction date approached, a manual operation was set up to make the first subassemblies for Spectra production. Manual workstations were set up both on "the beam" and in a separate location in the viewfinder assembly area. Jigs and fixtures assisted manual assembly. As many as 40 operators were working in the area on one shift during this manual phase of the project. Thirty-three of the operators were regular employees of Polaroid, in anticipation of an eventual three-shift operation requiring 11 operators per shift. The seven other employees, however, were temporary hires who were expected to leave once the peak need for extra hands had passed.

When pieces of automation were ready, they were installed. By mid-1986 the line was partly automated, with four stations of pick-and-place mechanisms and seven manual stations. At this point, three-shift operation began. Robot installations began in October 1986 and were completed in December of the same year. Commemorative plaques from November and December of 1986 are on display in the Optical Manufacturing plant to mark the dates on which each shift met its production quotas on the fully automated line.

Not all stations on the line met the reliability requirements necessary to satisfy the target production goals. The walking beam itself required major modification. Three of the pick-and-place devices were replaced by robots. The robot controllers, each in a box about 2′ × 2½′ × 2½′ high attached to the front of the line, were found to be obstacles to the operators. They were relocated to one side of the assembly room and connected to the machine by overhead cables.

On October 12, 1987, the Camera Division, which had been operating the line for approximately two years, formally accepted the viewfinder assembly machine from EFED, thus bringing the project to an end. The quality of the parts produced on the line at that point was sufficient to permit reduction from 100 percent inspection to sampling. The system had been designed to automatically reject parts that had missed an assembly operation, and the number of these rejects had also been reduced to an acceptable level by the time of sign-off. Further modifications to the line were expected, however, even at the time of this study.

HUMAN FACTORS CONSIDERATIONS DURING MACHINE DESIGN

Company Policies

Polaroid's policies regarding personal safety in machine designs were clear, explicit, and comprehensive. These policies were generally understood by all on the design project.

There were no other stated or perceived policies regarding human factors in the machine design stage. One general objective that guided the viewfinder machine design was to build the process to be as automatic as possible, minimizing the need for, or reliance on, workers.

The company's permanent employment policy, however, did influence the decision to automate the viewfinder assembly process. A manual assembly process would have required 130 to 140 new employees. It was anticipated that when the viewfinder was no longer produced, after 6 or 7 years, the company would not be able to absorb all these workers. To avoid either the burden of excess employment or violating their permanent employment policy by laying off these workers, EFED and Camera Division management jointly decided on automation.

The viewfinder assembly machine was planned as nearly complete automation. Workers would be required only to load piece parts into each assembly module and unload the finished product. Technicians would oversee the process and respond to malfunctions in the line. Technician involvement was anticipated as only contingent and infrequent.

The jobs required to attend to the machine were expected to be low-skill and low-pay-grade positions, with high turnover anticipated as a consequence. The design engineers reported that they expected operators would have little training, and the better operators would "bid out" to higher-grade jobs. For this reason, a design rule on the project was to keep decision making away from operators and to minimize permissible adjustments by operators. The equipment was also designed in conformance with a policy not to allow operators to use tools for equipment or process adjustment (e.g., operators were not permitted to have or use screwdrivers or wrenches).

Safety

Polaroid Corporation had a companywide safety program that applied equally to operating and staff functions. The program, headed by a corporate-level Safety Office, was unusually comprehensive. Not only did the company's activities cover a wide range of processes, from chemical manufacture and coatings to instrument assembly, but the company designed and made much of its production equipment. Its exposure to possible liability for injury or illness could have been great. Instead, the company had had an outstanding safety record for many years.

The Equipment and Facilities Engineering Division had its own safety engineering office, headed by a Principal Engineer whose prime responsibility was to see that safety was considered at all stages of equipment design. The EFED safety program was characterized by a certain degree of formality. Not only did it have a professional engineer dedicated to the program, but it had:

1. Mandatory forms that had to be filled out and signed off by the designer, the customer, and the safety officer. One form, the "Conceptual-Stage Coversheet and Checklist," was made out before the project was begun. It accompanied the request for capital authorization to top management. A second form, the "Safety Deviation Form," was prepared during equipment debugging to obtain authorization to operate a machine temporarily with known safety hazards that were to be corrected later. A third form, the "Safety Review/Acceptance Form," was prepared at the time responsibility for machine safety passed from the EFED to the operating division.
2. A computerized engineering design safety manual, so that all rules for equipment safety were immediately available by computer terminal to every design engineer. The easy-to-use, menu-driven program was installed on a VAX computer. It had taken the safety engineer four years to complete the installation, with sections on mechanical design, electrical design, and construction.
3. Extensive computer-based records. Records of the history of noise exposure of individuals working in high-noise environments, for example, were used in instances where there could be questions of work-related hearing loss.
4. Safety standards that, according to the EFED safety engineer, met or exceeded the following standards: OSHA, American National Standards Institute (ANSI), Factory Mutual guidelines, Massachusetts Department of Environmental Quality Engineering (DEQE), and the Joint International Code.

At the same time that funds were requested to design and build a new piece of equipment, the cost of safety equipment, including peripherals such as fire extinguishers, emergency shut-offs, nonskid floor treatment, or eyewash stations, was included as an integral part of equipment cost. According to the safety engineer, this reduced the possibility that certain safety items might be "finessed" if capital funds later became scarce. This policy also reflected the EFED rule that the equipment designer was responsible for the safety of the total environment of the machine, and for providing peripherals, such as those mentioned above.

The EFED safety engineer monitored progress on an equipment project during design, construction, installation, and debugging. While the machine was being installed, the safety engineer met with the manufacturing plant's Safety Committee (a combination of safety professionals, manufacturing engineers, supervisors, production workers) to provide information about the forthcoming equipment. He also conducted a review of the materials to be used in the product or process, to detect any hazard—toxicity, flammability, dust, fumes, and so forth—that might be present.

Design and installation of guards on a new machine was done after the machine had been installed and debugged in the plant. The safety engineer explained that engineers and technicians are permitted by law to operate equipment without guards, and that it was necessary to have all the necessary wiring and piping in place before guard designs could be made. He further stated that guards must look as if they are an integral part of the machine. "You can't put on a guard that looks like an afterthought."

Polaroid employed a contractor who specialized in constructing clear plastic guards for operator–machine interfaces. The preferred installation was a fixed guard requiring a tool to remove. This was not possible when, as in the case of the viewfinder assembly machine, machine jams were so frequent that fixed guards would have been impossible. Removable guards, however, had to have electrical interlocks that prevented machine motion when the guard was absent.

Once the equipment had been installed and run for a short while, the EFED safety engineer conducted a safety review, looking for the "worst credible incident" that might occur. The purpose of this review was not only to remove evident hazards but also to develop plans on how to "get the person out" in an emergency.

Operator safety was also a concern of the manufacturing division. A Job Safety Analysis was carried out for each job classification of production worker associated with a new machine, for a change of conditions on an older machine, or for a change of work environment. This analysis was done by the person's supervisor and the production worker(s) affected.

The EFED safety engineer was frequently called in by a plant to "qualify" supervisors in training operators for safe operation of equipment. He was also called in when there was an accident.

Documents shown to the case study group indicated that Polaroid's safety policies and procedures were followed in the design and installation of the viewfinder assembly machine. Sign-offs on the necessary forms had been obtained from the Camera Division. Concern about robot performance had led designers and manufacturing people to visit plants of other companies to see robots in action and to discuss problems. Guards had been installed over all sections of the line where moving parts were exposed. Clear plastic enclosures were built over the robots, assembly mechanisms, walking beam, and transfer sections. Doors and covers had safety interlocks and manual shut-offs, and emergency stop buttons were located at each module.

When observing the machine in operation, the study team found that parts jamming in the machine was a very frequent occurrence, continuing throughout the shift. Operators and technicians were required to remove guards, clear out offending pieces, replace guards, and restart that section of the machine. If the jam took awhile to clear, the workstations ahead of the module that was down would have to stop also.

It was found that operators and technicians had worked out ways in some cases of fooling the interlocks so machines could operate while guards were removed. The frequency of jams at one location on the machine was so high that the manufacturing group had arranged to have a window cut into the guard so an operator could reach in to clear jams without stopping the machine.

Despite the obvious concern for safety both in design and manufacturing, risks of injury were present in various parts of the machine. Examples included awkward "reaching in" to adjust a rotary parts feeder, the danger of having a hand caught between the walking beam and its guard when the operator fed housings into the line, and high overhead reaching to load magazines of lenses into the machine. Safety interlocks were installed to stop each module, but these did not affect the operation of adjacent modules, since each module was designed to operate independently. Signs had been posted to remind workers to shut down adjacent stations manually when opening the safety door of a module.

The safety record for the machine, however, was very good. The only accident of any consequence that was reported to the study group occurred to an operator working at a rotary transfer table. Her finger was cut when the machine cycled while a jam was being cleared.

Worker Comfort and Stress

Although EFED designers were clearly responsible for worker safety in their designs, it was equally clear that they were not responsible for providing operator comfort or for minimizing operator stress while working at a machine. Features such as properly designed chairs, floor mats, or lighting that would make an operator's job more comfortable were considered to be the responsibility of the customer's industrial engineering organization. When asked about the nonadjustable chairs, some with broken backs, that had been seen in the viewfinder assembly area, the EFED safety engineer stated, "The company has a problem with industrial engineers, they don't know how to use them."

Similarly, worker stress caused by a machine was not considered a responsibility of EFED. Workers experiencing stress were handled after the fact: "We have people [counselors] who listen to employees' problems."

Employee Involvement in Design

As was indicated earlier, there was no involvement of operators or technicians in the design of the machine. The work crew was assembled only after the system was installed, and no other Camera Division operators or technicians participated during the design phase.

Operator Control

The system was designed to minimize operator control. The nature of the product essentially mandated that the process be a hands-off operation. The primary function of operators was to be parts loading.

Feedback

The system included a computer that printed out data on key performance measures at the end of each shift, and this information was available to the operators. There were indications that the operators were aware of the results, and would leave work at the end of the day with a good feeling when the numbers were favorable.

Skills

The design intent was that the system would require minimal operator and technician skills. As is noted in the following section on impacts, on-the-job development of skills became necessary to keep the system functioning properly. This included use of maintenance tools on the part of operators and some elementary robot programming on the part of the technicians.

IMPACTS OF THE MACHINE DESIGN

The process of creating a new kind of assembly machine within the Equipment and Facilities Engineering Division, and its subsequent installation in the Optics Manufacturing plant in the Camera Division, had certain consequences for a variety of people within the company. In this section we report our observations of the impacts on design engineers, on manufacturing management and supervision, and on the operators working with the machine. The observations derive from discussions with individuals who were involved during the design phase and with managers, support personnel, and operators involved with the machine at the time of the study.

Impacts on the Process Designers and Design Management of EFED

The design of the integrated, robotic viewfinder assembly machine was an important step for the machine designers in EFED, but it was not without penalty for the division.

One of the most positive aspects of the project for the division was that the engineers learned how to design very sophisticated robotic equipment. This was not the first robot application in Polaroid, but it was the first attempt at multirobot integration. Many of the engineers had been fearful of the step. Now they could say with some confidence, as one of the engineering managers told us, "When the machine was completed, we were among the world leaders in small piece-part assembly using robots." There was a feeling of technical accomplishment within the division.

From a technical viewpoint, the EFED accomplished what Polaroid's top management intended. Not only did the machine produce high-quality viewfinder subassemblies in volume and at acceptable cost, but the designers also had mastered the art of robotic machine design. This learning fed back into the product-engineering group, which learned that cameras designed for automated assembly had to meet new design criteria. In contrast to designs for manual assembly, for example, parts destined for automated assembly had to be capable of easy and unambiguous orientation, and of relatively uncomplicated insertion maneuvers during assembly.

The designers learned that they could not count on someone outside the firm coming in and programming the robots for them: "You have to bite the bullet and do the programming yourself." It turned out that they actually had to teach vendors that were supplying parts of the machine.

The designers learned that designing the machine in modular stations that could theoretically be moved out and repaired off-line was not as successful or helpful as they had expected. As each station became more complex and dedicated to one step of the process, modularity had less and less meaning. The engineering project manager stated that if they were to design the machine again, they would "tie the machines together loosely" and not maintain the modular approach.

The designers also gained greater appreciation of the need for consistent quality of parts. Too much part-to-part variation aggravated the tendency of the machine to jam and caused significant productivity losses. Several of the parts, both from sources outside Polaroid and from within, gave the engineers great difficulty and were continuing to cause problems right up to the time of the case study.

During the viewfinder machine design and installation, the EFED had to establish and maintain interfaces with a number of organizations in Polaroid. These included Camera Engineering, which was responsible for Spectra design; Optics Engineering, responsible for the viewfinder optics and the critical dimensions of the viewfinder; the Program Office of Camera Division, responsible for translating the camera design into manufacture; and Optical Manufacturing, the plant that eventually was to receive and operate the machine. Some of these interfaces were very strong and successful. Others, notably between two of the engineering groups, were weak, and had to be compensated for by indirect means, by going through a third party.

Although it was not a novel experience for the Polaroid machine designers, this project did serve to reemphasize the importance of letting machine problems surface during debugging and of not hiding or masking difficulties that would continue to cause trouble. Project heads were emphatic that, in as complex a system as this, it was virtually impossible to anticipate every possible cause of fault, so it was important to be responsive to the problems as they became visible. In this instance, it meant that design engineers and drafters actually supervised the running of the machine in production during the months immediately following its start-up and did not relinquish responsibility for the machine for two years after it was put into the factory.

The director of EFED stated a principle of preserving what he called the "design intent" through the cycle of design and implementation. Design intent emerges from the early stages of conceptualizing the machine, when the essential characteristics of the machine are agreed upon. His point was that it was important to bring the people who would ultimately be responsible for the machine "into the design environment at the design phase and have them track along." By "people," he was referring to manufacturing professionals—engineers, supervisors, or managers, not operators or technicians. His view was that, after the design intent is established in the conceptual period, "then the losses begin." Changes occur, compromises are made. After the machine has been installed and has been in operation for awhile, people begin making changes that "totally violate the design intent." It is important to maintain and transmit to the user the core ideas of the design, because the creative designers do not stay around to protect their machine after it has been accepted by the customer.

The consequence of the success of this design project was that EFED was able sucessfully to take another big step—that of contracting for several Japanese-designed integrated robotic assembly machines representing another quantum step forward in process sophistication. They were able to do that confidently. As one engineer stated, "We didn't have to go to the supplier, hat in hand, and ask, 'What can you build for us?' We told them what we wanted, and we got it."

Despite the fact that EFED achieved the technical goals of designing and producing a machine on time that made a superior product at acceptable cost, the Division paid a price for the experience. The final cost of the machine ($4, 900, 000) was about twice what they expected to pay for it, and this fact was not appreciated by corporate officers who were unable to judge the magnitude of the technical challenge. Those who could recognize the nature of the accomplishment, including the president of the company, were more understanding. The reputation of the Division within the company, however, was damaged by the magnitude of the overrun and the length of time it took to work out the bugs in the system.

Outside of Polaroid there was a certain amount of favorable publicity connected with the machine. Several magazine articles were written about it, an American Society of Mechanical Engineers (ASME) meeting was held to demonstrate it, and the machine was described in Polaroid's Annual Report as an example of the advanced technology being used to produce the Spectra System.

Impacts on the Camera Division

The case study included meetings with Camera Division personnel at various levels, from division manager to operator. At the conclusion of the study, the team met with a "vertical slice" of the Camera Division organization that had been involved with the viewfinder assembly machine. The purpose of the meeting was to report to the Camera Division our preliminary findings and to hear opinions regarding the machine and the project.

Managers and Supervisors

Although somewhat dismayed at the high cost of the machine, Camera Division management appeared pleased with the machine. They liked the machine because it was cost effective in the high volumes that were needed, and because it enabled them to make clean assemblies. They also saw the machine as forcing a discipline with respect to parts quality and availability that was advantageous.

The division manager stated that he would "give the viewfinder project high marks," that it had been a difficult job with ambitious goals, and that good work had been done. He saw the project as having involved a fair amount of team building and creativity. He felt that one result of the project was generally increased competence in dealing with automation.

The division manager reviewed the history of process design policy at the Camera Division, explaining that in the early 1970s the division had set out to mechanize all aspects of camera assembly. A new camera design then being readied for market introduction had been regarded as the ultimate in instant-camera design, so its form was essentially to remain constant for a long time. Mechanization was seen as a means of dramatically increasing the quality of the product. In only three years after the camera was introduced, however, Polaroid decided to introduce a new, less expensive camera that was significantly different from the supposedly changeless design. At that point the division found it "had about 100 million dollars worth of boat anchors" in the dedicated machinery it had designed and built.

As a consequence of this experience, management policy had swung to the other extreme of designing cameras for very simple assembly, so manual assembly would be fast and easy. The change in philosophy was described as a shift away from "you design the camera; we'll find some way to make it," to the view that the product must be designed with ease of manufacturing in mind. In this view, as one person expressed it, design was actually a subset of manufacturing.

This approach was so successful that the division was able to reduce assembly direct labor content in the camera to 25 percent of what it had been in the highly mechanized era. The policy of design for ease of manufacture served the company well for over 15 years.

Now the division was seeking a compromise position in which assembly of certain camera functions was to be automated, while others remained manual. Earlier experience, however, had taught them to be cautious about dedicated machinery. The viewfinder machine was considered one of the early steps toward this new balance in process design.

Division management did not consider the new forms of assembly automation as being justified by direct labor savings. Instead, these systems were sold to top management on the basis of product consistency and quality, reduction of indirect (overhead) personnel, and job redesign. Camera Division managers told us, "We don't have a direct labor cost problem."

Impacts on Camera Division Technical Personnel

The manufacturing engineers and other technical support people we interviewed also liked the new machine, but they also saw limitations. The machine had clearly proved itself in terms of speed and capacity. To do the same job with hand operations would have been very costly in terms of labor, low yields, and lower quality.

These support people were still working to improve machine performance. Original productivity goals had not yet been met, and certain operations, such as removal of polypropylene film from the lenses, were not 100 percent reliable. The manufacturing engineers saw their task as being made easier by the modular design of the machine, which made it possible to pinpoint the sources of trouble. They could isolate a given function and work on it without interference from or impact on other interdependent functions.

Ease of maintenance and housekeeping remained a problem. One engineer suggested that, because operators had not been involved in the design phase of the project, some of the incipient maintenance and housekeeping problems had gone undetected until too late.

When asked whether the machine was sufficiently flexible in design to be used later for other camera assembly functions, the Camera Division engineers expressed the opinion that the machine's flexibility was quite limited. They tended to talk more about salvaging certain components of the machine (robots, some of the modules) rather than reprogramming or retooling the complete machine. The likelihood of technical obsolescence was given as another reason why the machine might have limited future use. By the time the machine had completed its mission for the present two cameras, many of the components, including the robots, were likely to be outmoded. It would be more efficient to acquire new equipment than to convert the old.

Impacts on Camera Division Operators

The design of the assembly process affected the operators in terms of the immediate nature of their work and the longer-range effects on employment, job advancement, and skill development.

As noted earlier, the design of a fully automatic process with expectations of minimal necessary operator intervention led to a design that locked out operators. When the machine was put into operation, therefore, the constant operator intervention and attention that was needed was made more difficult because the equipment had been designed on the premise that active operator involvement would not be required. Although formal company policy prohibited safety compromises, in practice operators were faced with a choice between following procedures that would result in unacceptably low production rates or intervening in the assembly process and putting themselves at risk by circumventing safety devices. The lack of ergonomic considerations also contributed to suboptimal physical and environmental conditions.

The impact of the project on operator employment can be viewed as having two phases. At first, a large number of temporary employees were hired to perform manual assembly tasks while only parts of the machine were working. Then, as the machine came on-line, these people were released. In the second phase, continued refinements of the machine finally resulted in performance at a level where the three inspectors, who were part of the regular production team, were no longer needed and were reassigned to other jobs.

In terms of job structure and longer-term quality of work life, the assembly machine did not appear to afford notable positive opportunities for the operators. The operator jobs were set up as unskilled, low-classification positions. There was a four- to six-step grade-level gap between loaders and technicians, which was too great to allow for advancement from loader to technician. A loader would have to transfer to a different production area to advance to the next higher grade level.

Because the formal machine design specified little technician responsibility for adjustment or maintenance of the assembly process, there was no initial recognition of the need for technician skills development. Actual informal on-the-job skills development was taking place, however.

The exigencies of actual assembly operation did result in changes from planned procedures. Some operators were given responsibility for minor adjustment and maintenance tasks, reversing a policy prohibiting their possession and use of maintenance tools. Some rudimentary training of technicians in teaching "points" to the robots had begun. The expectation was that the role of operators would be expanded, and additional training would be provided. Because these policy changes were just being enacted, it was not yet clear how they would affect operator advancement and compensation opportunities.

SUMMARY

The viewfinder assembly machine was a significant achievement for Polaroid's designers, who gained technical expertise and confidence in completing a very complex design. The Camera Division also acquired a better understanding of how to cope with advanced automation. The machine was only a qualified success, however. Actual performance of the machine fell short of expectations, even though production quotas were met. Machine jams, limitations in job growth, and stresses experienced by the assembly operators were symptoms of problems that might have been mitigated during design.

6

SIKORSKY AIRCRAFT SPINDLE/CUFF MACHINING CELL

The Company

At the time of our case study, Sikorsky Aircraft was the world's largest manufacturer of helicopters.* It was a division of United Technologies Corporation, an $18 billion company whose principal businesses were in the aerospace, defense, building systems, and industrial systems sectors. Other major divisions of the corporation included Pratt & Whitney (jet engines), Hamilton Standard (controls and propellers), Norden (military electronics), Carrier (heating, ventilating, and air conditioning equipment), Otis (elevators and escalators), and UT Automotive (automotive components).

Sikorsky, whose headquarters and main manufacturing complex were in Stratford, Connecticut, had a 1988 sales volume of $1.7 billion. The majority of sales were to U.S. military customers, but international sales of military and civilian helicopters constituted an important fraction of total volume. Pretax operating profits of the division were not publicly available, but the combined profits for the Flight Systems business unit (comprising Sikorsky, Norden, and Hamilton Standard) were $97 million on sales of $3.2 billion in 1987 and a loss of $15 million on sales of $3.5 billion in 1988. The loss was due to a $120 million restructuring charge within the Norden division.

Sikorsky sales in 1986 and 1987 were $1.6 and $1.5 billion, respectively, reflecting a somewhat lower level of U.S. military procurement than had been experienced in prior years (Table 6–1). The division's major competitors were Aerospatiale (France), Bell-Textron (U.S.), Boeing Helicopter (U.S.), and McDonnell Douglas (U.S.).

*Dates of on-site visits: March 1, 2, 9, 10, and 27, 1989.

Table 6–1 Sikorsky Annual Helicopter Shipments (units)

1981	1984	1985	1986	1987
189	211	201	159	164

Source: United Technologies 1988 Fact Book

United Technologies' worldwide employment in 1988 was 186,800, down from a peak of 193,500 in 1986. Sikorsky's total employment in 1988 was approximately 13,000, of whom 12,000 were in Stratford. The factory work force was unionized, represented by a Teamsters local.

Labor relations had been relatively stable for a number of years. Pay levels of employees were described as being at or slightly above prevailing community wages for comparable work. Many members of the division were long-service employees. The distribution of ages of employees at the Stratford location was bimodal, with an older 50–60 age group and a younger 25–35 age group.

The main Sikorsky plant, which incorporated division headquarters, product design, manufacturing engineering, parts fabrication, assembly, hangar, and flight test facilities, was bounded on one side by the Housatonic River and on other sides by major highways and residential/commercial areas. Expansion at this location was consequently restricted to what they could accomplish within the available area. This made space utilization in the factory an important strategic factor.

Capacity limitations had forced Sikorsky to subcontract much of the fabrication of parts to other firms. Such subcontracting tended to increase Sikorsky's manufacturing costs. If it were possible to increase capacity through improved technology or methods, the division's competititive strategy would be to bring such work back into its own shop. This case study centers around the design and installation of a multistation automated machining cell that accomplished that objective.

THE PRODUCT

Sikorsky Aircraft designed and produced military and commercial helicopters. Its 50-year history in this business began with Igor Sikorsky's initial demonstration aircraft flight in 1939. Although helicopters saw limited use in World War II, both military and civilian applications began expanding rapidly in the 1950s. In the limited-scope wars in Korea and Vietnam, helicopters became key tactical weapons for attack, support, and rescue. Commercial applications expanded, also, in construction, transport, and oil and gas operations. Sikorsky's progress in product development and performance was maintained throughout.

The helicopters in Sikorsky's product line were primarily medium to heavy models, weighing from 16,000 to 73,000 pounds.Typical of the medium-weight aircraft were the UH-60A Black Hawk, the UH-60B Seahawk, and several other derivative models within the UH-60 series. Built primarily as a combat helicopter, the UH-60 series served in missions that ranged from transport of combat troops (one could carry a squad of 11 infantrymen) to antisubmarine warfare. The Marine Corps CH53E Super Stallion and its Navy counterpart, the MH-53E Sea Dragon, represented the largest of Sikorsky's line, and were the largest helicopters in the Western world. They were capable of carrying 55 combat troops and could lift 16 tons of equipment. In addition to their transport capability, these craft were used for mine countermeasures and rescue operations.

Sikorsky also made an intermediate-weight helicopter, the S-76, designed for commercial operations. During the 1970s and 1980s, these aircraft provided ferry service for crews and equipment to offshore oil-drilling platforms and to other remote sites. The S-76 was also used for executive transport and for air-rescue missions. One derivative version, the H-76, was used by the military in a variety of roles.

Although the product line in 1989 included 18 models of helicopters, almost all were derivatives of three basic aircraft designs: the UH-60A, the CH-53E, and the S-76. Modifications were made to these generic designs to adapt the aircraft for different missions and environments. Increasing use of helicopters for surveillance and other technologically sophisticated roles had increased the amount and complexity of electronics to the point where over half the cost of a military helicopter would be in its avionics systems.

A successful new product design, such as the UH-60A Black Hawk and its derivatives could expect to be in production for 20 years or more. Thus, although there was a continuing flow of modifications and improvements, the basic structure and drive mechanisms were likely to be manufactured for a long time.

Even so, production quantities were not large. During the period from 1983 to 1987 an average of 137 units of the highly successful UH-60 series was produced annually. Total cumulative production of this class of helicopter, which was introduced in 1978, was expected to reach 2253 by the year 2007, according to United Technologies' Annual Report for 1988. Production of the much larger CH-53E and its derivatives averaged 14 per year over the 1983 to 1987 period.

Despite its dominant position in the helicopter market, Sikorsky found its competition to be strong. Both performance and price were significant factors in the procurement decisions made by its customers. Adding to the uncertainty always present in a competitive situation was the fact that the U.S. military was soliciting competitive bids for helicopter spare parts. The helicopter replacement-parts market represented a significant portion of Sikorsky's annual sales volume. This move by the military meant that in the replacement-parts business, too, there was competition. By subcontracting parts fabrication, then, Sikorsky made itself more vulnerable to competition in spare parts. Competition could come not just from other aircraft manufacturers but from Sikorsky's own subcontractors, who had learned how to fabricate the parts when supplying them to Sikorsky. So subcontracting carried with it the risk of a subsequent loss of the profitable spare-parts business.

Because human lives were totally dependent on the faultless performance of many critical components in a helicopter, parts quality received careful attention at all stages of manufacture, from raw materials inspection through final assembly and test. The complete history of a critical part could be traced through records that were maintained by the division. The need for close control of quality provided further motivation to keep manufacture of critical components "close to home."

Among the machined metal components critical to the reliability of the helicopters were the "spindle" and "cuff"—parts that joined each rotor blade to the power train of the aircraft. The machining cell that is the focal point for this case study was designed to machine the spindles and cuffs for the UH-60 series and S-76 series of helicopters. These parts were machined from titanium forgings weighing up to 90 pounds in the raw forging state. During their transformation into finished parts, the forgings were turned, milled, ground, deburred, drilled, profiled, bored, and inspected (see Figures 6–1 and 6–2). Many of the machined parts were very high-cost items. This was true both for the forging itself and for the labor to machine and process the part. Damage to a part during or following machining could render the part useless. Any distortion of the threaded portion of the spindle, for example, would be cause to reject the entire part.

Figure 6–1 Black Hawk Spindle (machined, right; forging, left)

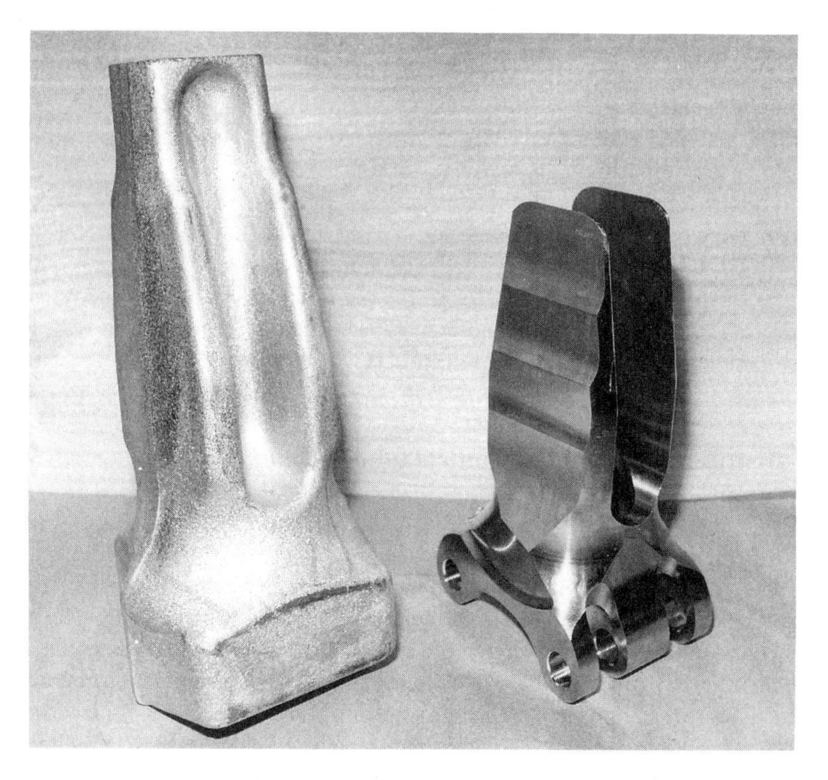

Figure 6–2 Black Hawk Cuff (machined, right; forging, left)

THE PROCESS: SPINDLE/CUFF CELL

Sikorsky's main machine shop was part of a production plant that included fabrication of sheet metal parts, construction of composite rotor blades, machining of helicopter parts, and assembly of finished helicopters. The machine shop occupied 162,000 square feet of the 2-million-square-foot plant and included about 250 metal cutting and grinding machines.

The machines were arranged in two sections, a general machining section and a gear section. In the general machining section there were about 70 Distributed Numerical Control (DNC) workstations connected to the DNC System described in the next section and 150 conventional machines including Numerical Control (NC) lathes for cutting and grinding general helicopter parts. In the gear shop there were about 40 machines: conventional parallel axis and bevel gear cutting and grinding machines as well as NC lathes. Approximately half of the shop's output came from DNC machines and half from the conventional machines. The floor of the machine shop, made of small wooden blocks with the grain facing up, was convenient for rearrangements of machines and wiring, was safe in case of oil spills, and was easily replaced and cleaned.

The machine shop had been in transition from a traditional functional arrangement of machines to one that incorporated both Group Technology and Computer-Integrated Manufacturing. A large number of the machines had been grouped in cells that were equipped with automated material handling systems for transferring parts between different machining operations.

There were four cells, each dedicated to a "family" of parts: (1) spindle/cuff cell, (2) flange cell, (3) small housing cell (FMS), and (4) hub cell.

Advanced group technology for production at Sikorsky was first introduced in 1981. At the time of the case study, the technique was well established and was being expanded. The objective was to increase capacity and to minimize moves of processed parts, thus cutting down work-in-process inventory.

Distributed Numerical Control with Machine Tool Monitoring (DNC/MTM)

The machining processes for all of the NC machines in the shop were administered by a central computer system through a Distributed Numerical Control Shop-Floor Control System that delivered work orders to the individual machines and monitored their performance. The DNC/MTM system was a communication network for the machine workstations and for manufacturing, management, and engineering departments. It established a common information base for all activities, including production, engineering, finance, and administration.

The DNC/MTM system was made of units arranged as a cascade in the following order:

1. A host IBM 3090 mainframe computer, which was linked by an SNA gateway to a VAX 11/785 and a VAX 8500 computer, coupled together and connected by Ethernet cable to machines, supervisory stations, and other support stations on the shop floor.All three computers were located in an Information Management Data Center.
2. DEC RT-103 workstations in the machine shop, equipped with NC and monitoring sensors and linked to the VAX cluster.
3. DEC VT-220 terminals located in the machine shop for labor and scheduling data, also linked to the VAX cluster.

The VAX cluster was an autonomous unit that acted independently of the IBM host. However the production database was stored in both the IBM and VAX computers, creating a double backup of data. The VAX cluster itself, by the mere fact of a parallel connection, provided redundancy; that is, a breakdown in one of the two units would not disrupt the flow of information.

The software used by the computers included the following:

The IBM host:

1. CADAM INC—Computer-Aided Design and Manufacturing, a CAD package developed by Lockheed Corp.
2. CATIA—Computer-Aided Three-dimensional Interactive Applications, a computer modeling software package developed by Dassault of France.
3. OPT—Optimized Production Technology, a finite scheduling software package run on the IBM host and transferred to the VAX cluster for work order requirements and routing. Developed by Creative Output Ltd, of Israel.
4. APT-AC—Automatically Programmed Tool, software for advanced contour programming.

The VAX computers:

1. NCAM—Numerical Control and Monitoring, a software package developed by GE Fanuc Automation Corporation, that stored and transferred MCD (tape data) to and from the workstations.
2. SIMS—Shop Information and Management System, for labor reporting and part-scheduling data.

The workstations had a direct communication link with the VAX cluster: the input from VAX included NC machine production instructions, and the output included feedback from the machines on production and tool performance. The workstations had CAM graphics with full production documentation and zoom capability. Some workstations accepted individual machining programs prepared by the machine operators.

How the DNC System Worked

The flow of information for production started from the top, at the IBM host, and ended at the workstations. Design and drafting was performed by the Engineering Department on workstations linked to the IBM computer, and the resultant information was sent to the VAX cluster in the Manufacturing Department. There this input was used for working out the data that were transmitted to the workstations in the form of work orders; schedules; production programs; requests for fixtures, cutting tools and gauges; and tool paths.

A countercurrent flow of information, from the operators and tool sensors at the workstations on the floor, went upward to the VAX cluster. The information included feedback on production runs, confirmation of part completion, inventory, and tool performance and wear.

The two-way information flow provided the needed control of the production process. The system not only was able to optimize the performance of machine tools, but it also helped to speed the production flow, reducing the need for storage space for parts in production.

In addition to the above, there was a flow of in-process inspection data for quality control. The data from the VAX cluster and the workstations went into Statistical Process Control (SPC) software analysis. The SPC software performed automatic data analysis and fed back graphical information that allowed the operator to control the process. The DNC system served as a common database for product quality information.

Spindle/Cuff Cell

The spindle/cuff cell that was chosen for case study was a product of in-house design and development, the fruit of Sikorsky's manufacturing staff efforts. Its conception and original design were created by a team of Sikorsky engineers, manufacturing personnel, and other support groups that were to be involved with the final product. For the most part, the machines in the cell were assembled from the complement of equipment already in the shop. The automated storage and retrieval system, automatic guided vehicle, and a few of the machine tools were new.

The cell included 21 metal-cutting and -grinding machines and two transportation systems: an automated storage and retrieval system (ASRS) and an automatic guided vehicle (AGV). It also contained a one-person inspection station, providing additional quality control. Its main products were Black Hawk helicopter spindles and cuffs, although it was also used for production of other parts of similar size and complexity. The cell was located in the main section of the machine shop.

The cell comprised the following machines:

	Units	Manufacturer
Four-axis, single-spindle machining center	4	Excello/Mazak
Three-axis, four-spindle machining center	4	Rambaudi
Three-axis, three-spindle machining center	2	Cincinnati Milacron
Milling	2	Hypowermatic
Lathes 18″	2	Cincinnati Milacron
Lathes 10.5″	2	Cincinnati Milacron
Radial drills (conventional)	2	Carlton/Invema
Surface grinder (conventional)	1	Danobat
OD grinder (conventional)	1	Landis
Saw (conventional)	1	Wilton
total	21	

All but the conventional machines were linked to the DNC/MTM system. Operators on the conventional machines followed work orders and schedules prescribed by the DNC system and used RT-103 terminals at their machines to feed back labor and schedule information.

The machines and the ASRS were arranged in four parallel rows, running south to north, in the following order (see Figure 6–3):

First row: four lathes and two 3-axis machining centers. Second row: ASRS and the supervisor's station. Third row: four 4-axis machining centers and one mill-

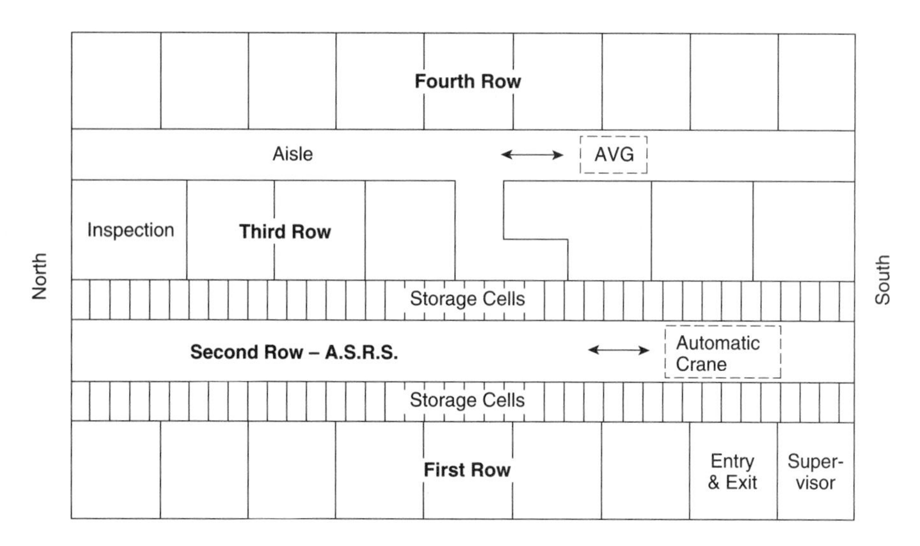

Figure 6–3 Spindle/Cuff cell

ing center, a radial drill, and an inspection station. Fourth row: three 3-axis machining centers and six conventional machines.

Control

Two independent computer units were in command of the cell: one, the VAX cluster described above, and the other, a local Digital Equipment PDP 11/44 computer.

The PDP computer gave commands to the ASRS and AGV to control the supply and retrieval of parts from each of the 21 machines. The PDP was manned on only the first shift. Its operator's main activity was to correlate the supply and retrieval of parts according to the work schedule provided by the DNC. There was a supervisor at the station next to this unit on the second and third shifts. The supervisor would correct any minor problems that might occur. The system ran with minimal human interaction on these shifts.

The manufacture of a typical spindle or cuff going through the system required about 20 to 30 machining operations. Completion of a part took about 10 to 12 weeks as compared to 20 to 25 weeks before the introduction of the ASRS.

Automated Storage and Retrieval System (ASRS)

The ASRS was a "first-in, first-out" storage system. It was fully automatic, controlled by the PDP computer. Storage cells in the multitiered steel rack were loaded and unloaded by a computer-controlled stacker crane. The system was built and installed by the ACCO Material Handling Group of Babcock Industries Inc.

The ASRS stacker crane supplied parts to the machines in the first and third row and retrieved parts after operations were completed. Machines in the fourth row, which could not be reached by ASRS directly, were serviced by an automatic guided vehicle. The AGV ran along a corridor between the third and fourth rows (see Figure 6–3) delivering parts from the ASRS. By controlling the supply of parts, the ASRS controlled the production flow for the entire cell.

Because the ASRS accomplished more than simple movements of parts and tools, the designers of the system stated that ASRS in this instance really meant "Automatic Scheduling and Routing System." The ASRS, guided by intelligence from the computer control system, acted as the heart of the cell. It furnished parts and tools to each machine, when needed; retrieved them when an operation was finished; and provided orderly, protected, storage until they were scheduled for further work.

The ASRS was a steel-frame storage structure about 131 feet long, 15 feet wide, and 12 feet high. It had about 200 storage cells, arranged in two parallel rows, and a number of loading docks. Each storage cell contained one portable open container for storing parts or tools. An automatic crane moved back and forth in the aisle between the two cell rows, stopping for loading or retrieving. The loading docks located just outside the parallel cell rows delivered parts and tools to the machines or received them back. Each loading dock had a short roller conveyor that moved containers directly to a machine area. The docks matched the locations of the individual machines—in general, one dock per machine. There were also loading docks for the AGV and the inspection station, as well as a manned dock for entering and removing items from the system.

Automatic Guided Vehicle (AGV)

As mentioned above, the automatic guided vehicle served the rows of machines the ASRS could not reach. Its dimensions were about 8 feet long, 4 feet wide, and 4 feet high, and it was also installed by ACCO. The vehicle was fully automated, guided by a wire embedded in the plant floor, and equipped with sonar and a pressure-sensitive bumper. It moved on a preset path, the aisle between the third and fourth row of machines, with "side streets" to the ASRS loading dock and to the individual machines. It stopped whenever its sonar sensed an obstacle and resumed moving when the obstacle was gone. Its speed equaled that of a slow pedestrian.

At the time of our visit the capacity of the vehicle was not being fully utilized, and plans were made to give it other routes as well.

THE DESIGN PROCESS

Project Initiation

The Spindle/Cuff Machining cell was Sikorsky's third group-technology project. It was first conceived in 1981 when it became apparent that in-house capacity for fabricating spindles and cuffs could not keep pace with a rising demand for helicopters. To meet the projected demand, the company faced the prospect of subcontracting these parts to vendors at a premium cost.

At the time, spindles and cuffs were being machined in functional sections of the shop, which meant that the parts-in-process moved around the entire machine shop. Parts on pallets were moved manually between units and were transferred from pallet to machine by crane. Setup costs and transportation delays resulted in a large in-process inventory. Often, parts were damaged due to handling because they were transported back and forth across the shop several times. For the number of spindles and cuffs that Sikorsky had to produce every month, this method of manufacturing was inefficient.

Outside contracting offered no better alternative. These parts were critical components in the helicopter, and sending the work outside meant that the company had greater difficulty controlling quality. Costs tended to be higher. With subcontracting there was also the likelihood that a vendor would become a competitor in the lucrative spare-parts market. The in-house group-technology approach with materials handling automation, therefore, promised improvements in cost, quality, and inventory and reduced the threat of competition for future business.

A project team was formed in 1981 with the goal of designing a manufacturing system that would make more parts with the same amount of space and people. The core of the design team consisted of a manufacturing engineer, a facilities engineer, and an industrial engineer, but the total team included production supervision and machine operators as well.

The team had to justify cost savings to satisfy corporate ROI rates for new machinery required for the system. As was true for the Flange Machining Cell project that preceded it, the goal of the cell design project was to streamline production flow and to increase capacity by utilizing existing labor and machines more efficiently. The designers were instructed to (1) increase production capacity to 100 spindles and cuffs per month from 30 per month, and (2) stay within available space on the floor. It took the team seven months to complete the conceptual design within the constraints that were placed upon it.

The team's design employed group technology. All the machines required to produce the parts were grouped in two parallel rows with an automated materials storage and retrieval system between them. The ASRS would transport tools and parts among the machines. The system would initially use 13 existing machines and 4 new machines in a configuration similar to that of Figure 6–3.

The Request for Capital Appropriation for the first phase of this equipment was submitted in October 1982. It mentioned the Material Handling System, but the funds for that were not requested until January 1983.

Design Project

During the design phase, the manufacturing engineer on the team prepared a presentation intended to involve shop-floor people. When the system concept had been developed, he created a storyboard with the proposed layout of the spindle/cuff cell and then presented the idea to foremen and operators on each shift. He also mounted the layout on a wall and encouraged foremen and operators to come in, look at the layout, and make suggestions. Several useful improvements came out of this process.

During 1982, the various machines making up the cell were assembled in their new configuration. Specifications for the automated material handling system were prepared by the Sikorsky team. They were complete at the end of 1982, and the system was put out for bids. A total of 14 vendors expressed an interest, but after examining the specifications, 13 of them withdrew from the bidding process. The remaining bidder, Acco Babcock (now known as Acco Systems), was awarded the contract.

A material handling system (dubbed the Automatic Scheduling and Routing System) was put into place. A system of this particular configuration had never been built before. Acco ended up writing much of the software for the system on the shop floor after the hardware had been installed. Considering the magnitude of the task, progress was rapid, and the system was installed and debugged by the middle of 1983. By the end of 1983 shop management signed off on the project. It had been completed on schedule.

Worker Reaction

Immediately following installation of the cell, a few of the workers in the area felt a certain amount of insecurity with respect to the new system. Some of them were afraid of plunging into an unfamiliar situation. Some of the older workers were not certain they would be able to master the new work methods, while some of the younger people felt they might get trapped in the same job. Although these people were the exception, their resistance continued for several weeks.

After some time, the operators were reassured that no one was going to lose his or her job, and things settled down. The operators started working as a team and gradually began to accept the system. Because there were now many different machine types in the same area, operators who had been specialists in one type of work began to express interest in learning how to operate machines other than the ones on which they had been trained. As the advantages of the new system became apparent, worker enthusiasm grew. In the end it was the operators who convinced management of the effectiveness of the new system.

In 1988, the spindle/cuff cell was expanded by the addition of a third row of machines. An automatic guided vehicle was installed to transport tools and work pieces between the ASRS and the machines. Up to this time the local PDP 11/44 minicomputer had been controlling the ASRS, but in 1988 the project team began tying the ASRS into the plant's central Distributed Numerical Control computer network. The completion of this phase was scheduled for the end of 1989.

Cell Performance—Productivity

Prior to integration of the machines to manufacture spindles and cuffs, the same products were fabricated by 33 machines scattered throughout the shop. The creation of the integrated spindle/cuff manufacturing system not only decreased the number of machines required to make the parts but improved characteristics of the products and the production process. Less setup time, computerized scheduling, minimal operator intervention, and distributed numerical control were some of the attributes that improved the performance of the cell.

The cell was designed for an expected throughput of 80 Black Hawk spindles and approximately 90 cuffs per month, plus 20 each of S-76 spindles and cuffs, 40 vibration absorbers, and 100 aluminum airframe fittings.

At the time of this study the line was producing Black Hawk spindles and cuffs at the planned rate, as was true of the rates for the other components. A three-shift work force of 46 was required for the total level of output, of which only 18 were required for the Black Hawk spindles and cuffs. According to the manufacturing engineers involved with the cell, a record production of 250 spindles in one month, approximately the peak capacity of the cell, had once been achieved, with production remaining at or near 200 per month for more than a five-month period.

The production output of Black Hawk spindles and cuffs before cellular manufacturing began was approximately 30 each per month, requiring a work force of 60. The throughput time came down from 20 weeks to 10 weeks. The cell had greater reliability and precision, which resulted in a substantial improvement in part quality. There was a 6-to-1 reduction in the rejection rate of parts made in the cell, as opposed to the previous in-house method of manufacture.

All of the machines in the cell came under a distributed numerical control (DNC) shop-floor control system. Not only did the operators get machining commands from the central computer, but they fed information about finished parts and labor performance into the DNC network. This continuous updating of manufacturing data enabled the computer to display queue lengths by part type or by machine type. This feature permitted much tighter scheduling of operations.

Smooth material flow within the cell decreased the amount of work-in-process because work pieces were not "lost" in transit. According to the Senior Industrial Engineer on the project:

> Even though we put multispindle milling machines in the spindle cell to increase capacity, the work-in-process decreased effectively. Previously the machines were all over the shop floor, and many times unfinished parts remained unclaimed or unnoticed. This increased the work-in-process. At present, manufacturing of spindles and cuffs remains restricted to the cell, and parts get transferred by an AGV.

System Losses and Setup Time

As a group, the utilization rate for the numerically controlled machine tools in the system was approximately 70 to 80 percent of their capacity. The availability (uptime) of the machines was quoted to be in the range of 92 to 96 percent. The target availability was 95 percent, which, according to the engineers, was hard to sustain.

Setup time on the machines was reduced by loading the work pieces on hydraulically actuated fixtures that had devices for checking accurate alignment. Most of the multispindle milling machines and all of the four-axis, single-spindle milling machines had extra fixtures mounted on the bed of the machine so operators could load the next set of work pieces while the machine was in operation.

The five conventional machines in the cell had comparatively higher idle times. The AGV was underutilized and was idle for more than 50 percent of the time. Not all machines were served by the AGV. Many received tools and jobs directly from the ASRS. At the time of the study, plans were being made to utilize the AGV to serve adjoining machines that were not part of the cell.

System Flexibility

Because it had been installed for the purpose of making a rather limited family of parts, the system was not required to demonstrate great flexibility. The range of prismatic and cylindrical machining was restricted to that required by spindles, cuffs, and a few other related parts. All machines, however, were general-purpose tools, so flexibility was only a matter of programming and fixturing, rather than of new equipment installation. Within the limits of its designated mission, the cell did possess limited flexibility:

Machine Flexibility: The machines in the cell had an efficient arrangement for changing fixtures that greatly reduced setup time. The newer milling centers had automatic tool changers. The machines could be reset with respect to the NC program quite easily, and machining instructions and code could be read by the operator through the DNC network.

Product and process flexibility: Approximately 15 different kinds of spindles, cuffs, and fittings were produced in the cell, but the variations among all of the spindles and cuffs produced were minor.

Routing flexibility: The system had the ability, in some instances, to route parts to alternate machines if a particular machine should happen to break down.

Volume flexibility: The system had shown a capability of operating at substantially higher rates of output under crash circumstances (up to 2.5 times "rated" capacity).

Maintenance

No maintenance work was carried out by operators. In case of machine breakdown, the operator reported the matter to the foreman. There were also means of informing the Maintenance Department through the DNC system, but normal practice was to inform the foreman first.

HUMAN FACTORS CONSIDERATIONS DURING SYSTEM DESIGN

Worker Involvement in Selection and Design of New Equipment

There was an implicit policy in Sikorsky's shop that employees at all levels affected by technological change were to be directly involved or consulted in the selection of new equipment and process changes. The involvement of operators was to occur before the selection of new equipment and, through an active suggestion system, during process implementation. This unwritten but generally understood policy was mandated by the Vice-President of Machining and Transmission Manufacturing. Operator involvement was not a part of previous corporate culture or practice; the suggestion system had existed earlier, but was not as widely publicized before the current Vice-President's tenure.

The extent of operator involvement varied in each project, but it was widely understood that operators should be actively involved in the final evaluation of proposed new equipment, and at least consulted in production system and process design. In the spindle/cuff machining cell, the shop-floor workers became involved in the selection process for the new machines and handling system after the initial capital appropriations request was approved in 1981. The foreman was a member of the committee that established the equipment specifications for each Request for Proposal (RFP) and reportedly consulted with operators during this phase of the process. When vendors submitted proposals, the committee selected the proposals that met the specifications of the RFP.

Operators were involved in evaluating the equipment and proposals of the qualifying vendors. Operators attended trade shows and visited other companies using each vendor's equipment. In some instances it was reported that operators even attended trade shows on their own time and at their own expense. The operators and the foreman then prepared a report that was sent to the Vice-President. Operators also met with other members of the equipment selection committee to discuss the different systems.

Human factors considerations in the actual design of a machine system were largely dependent on how outside suppliers designed their standard machines. If Sikorsky had wanted to incorporate changes that would have made greater use of human capabilities, it would have been difficult to achieve. Because the factory was essentially a large job shop, the company usually bought no more than one or two machines of a given type at one time. Consequently, they found that equipment manufacturers were not highly responsive to their suggestions for design modifications without substantial price increases. The principal option available to the in-house system project team was to buy from the tool builder whose design was closest to what was considered ideal. But even in the areas of safety, ergonomics, and comfort, managers and operators described examples of machine designs that exposed the operator to possible injury during particular operations, increased operator fatigue, or failed to provide for operator comfort or ease of operation.

Operator Suggestions for Postimplementation Process Improvement

Operators were encouraged to suggest improvements in all aspects of the process or plant (e.g., from changes in machines to the color of the walls). All suggestions were reviewed by the Vice-President of Machining and Transmission Manufacturing and then assigned by him to the appropriate office for evaluation. The person to whom the suggestion was referred was required to respond to the Vice-President's office within 30 days with either a completed evaluation or a status report (e.g., in the process of obtaining relevant information for an evaluation). Suggestions by hourly employees that resulted in cost savings were rewarded by cash payments ranging from several hundred dollars to as high as $5000. Suggestions that were implemented but did not produce clearly demonstrable financial savings were awarded at least $25.

Safety

Sikorsky's explicit safety policies were described as being at least as stringent as OSHA standards. Safety provisions on machines were largely the standard means provided by the equipment manufacturer. The operating areas of newer computer-controlled machines were entirely enclosed during operation. Older machines were fitted with local chip guards.

Two policies, an explicit policy requiring safety glasses or shields and an implicit policy for maintaining high levels of cleanliness, were the most visible policies requiring worker participation in safety conditions. The Vice-President emphasized the requirement that everyone on the shop floor wear eye protection.

An implicit policy to keep the shop floor clean and free of debris was a formalized procedure. There were automatic chip-disposal devices at most workstations, although on some machines the operator periodically had to sweep the chips onto a disposal conveyor belt. A floor-sweeping vehicle regularly patrolled the aisles, and aisles were scraped and resurfaced every three weeks. There appeared to be little debris that fell from the machines to the floors, and operators maintained the cleanliness of their areas. The shop was also subject to frequent inspection and tours by government officials and other customers, adding to the emphasis on maintaining a "presentable" appearance.

Overhead cranes were provided at each workstation for loading and unloading parts. The cranes were available to the operators, although it was expected that operators would manually lift parts weighing less than 35 pounds. Roller conveyors were generally located in convenient positions and heights for easy loading and unloading, although some reaching and bending was required at several stations.

Comfort

There was no explicit policy regarding comfort. Since there was little in-house equipment design, the degree of operator comfort at workstations was largely the result of design by the machine manufacturers. The implicit policy for operator involvement in equipment selection and for suggestions after implementation, however, allowed some measure of operator involvement in comfort considerations.

The wooden block floor, designed to facilitate changes in wiring and machine placement, provided a better surface than concrete for standing and for absorption of chips and spilled liquids such as oil and coolant. Platforms were provided, to adjust for the height of machine controls and loading stations. In response to an operator's suggestion, a platform was built on one machine that made it easier for the operator to load and unload parts from the ASRS.

Maintenance

The Maintenance Department did all repairs and routine maintenance. There was no formal provision for operator maintenance of the machines. For maintenance and repairs the operator had to obtain a foreman's authorization before submitting a request to the Maintenance Department. The request was then submitted to the Maintenance Department by telephone call. The DNC system did allow operators to send comments about requested repairs to supplement the telephone calls.

Operator Control and Modification of the Process

Throughout the shop the extent of operator autonomy and control over a process was dependent upon a combination of factors, including machine type, part type, supervisor policy, worker initiative, and shift. There was no formal policy, and actual practices varied. There was also a distinction between "programming" and "editing." Programming consisted of writing the instructions for the machining operations, and editing involved making modifications to a program that had already been written, whether by an operator or programmer. One class of editing consisted of changing stops, feeds, and/or tool alignment, or compensating for differences in the size or composition of the raw part materials. These changes did not alter the sequence or tool path. Another class of editing involved substantive changes in the program, changing the path of the tool or optimizing the process by devising shortcuts and other changes.

In the spindle/cuff area the machines were designed to allow operator editing but not substantial programming. There was a general rule that operators would not make any programming changes. Programming on many of these machines was possible, but not easily done. The designs of many parts were critical and machining programs were "frozen," prohibiting any changes without approval and certification. Parts also differed in complexity of machining and thus required different levels of skill to write the programs.

In another area there were operator-programmable machines. These machines were designed with full programming capabilities at shop-floor controls that allowed programming through use of a series of menus.

Learning and Skills

Company policy encouraged skills development in four different ways:

1. A company training program that prepared people for specific jobs. The company had a formal, state-certified training program in machining skills. Workers could enroll in the program at company expense.
2. Participation in the design of work cells and selection of equipment, which broadened operators' knowledge and judgment.
3. Operator programming of machines, where permitted, which further increased specific skills.
4. A general tuition reimbursement policy, which provided 75 percent of undergraduate tuition and 100 percent of graduate education tuition.

IMPACTS OF THE SYSTEM DESIGN

The introduction of any new technology tends to have multiple impacts on the internal organization.The design and implementation of the Spindle/Cuff Machining Cell was a major step for the Sikorsky shop, even though it was not the first group-technology cell introduced. It was, however, far more complex than the Flange Machining Cell that preceded it, and involved a significantly greater use of automation technology and flexible manufacturing techniques. There were consequences for managers, engineers, and operators. There were consequences, too, for the product and the customer.

Impact on Management

The senior manager of the shop, the Vice-President of Machining and Transmission Manufacturing, used the introduction of the DNC/MTM System to initiate a series of major changes that increased the capacity and performance of the shop. Once this system was installed, it was possible to group machines to make families of parts and to introduce automated scheduling, movement, and machining of these parts. The management essentially accomplished a bootstrap revolution in parts processing.

In addition to encouraging the technical change, the Vice-President also insisted on operator involvement in the design and selection process for the new technology. This approach had benefits for the operators that are described later, but it had definite advantages for management as well. Although unionized, Sikorsky experienced little or no resistance from the union to the changes that were taking place, as long as the operators were fully involved. When resistance was experienced, in mysterious machine stoppages shortly after the cell was installed, it was motivated by concern that jobs were being eliminated and by perceived loss of operator control over the selection of work. The resistance was short-lived, but served to indicate that even more involvement of operators during the change might have been beneficial.

Managers expressed the opinion that the team approach, which linked at an early stage all the people necessary for an acquisition, might be slower to arrive at initial decisions. Because it eliminated serial, or stepwise, processing of decisions, however, the overall time could actually be shortened. Managers considered the more knowledgeable operators to be "invaluable consultants—they can really pick apart a machine."

Perhaps the most important impact of this project on shop management was the realization that within the plant they had the requisite expertise to make the major changes that were needed. It was possible for them to achieve efficient use of machines, people, and time. The sense of confidence generated by this experience was reflected throughout the shop organization.

Impact on Manufacturing Engineers and Programmers

There were about 400 manufacturing engineers, industrial engineers, and programmers in the Manufacturing Department. Many of them came during the last decade directly from universities. Fewer people were rising from the shop level through the ranks without a degree.

The introduction of the DNC system, which had been installed over a large part of the shop and was connected to the spindle/cuff cell, had had a significant impact on the engineers and programmers of the department. Their work methods changed: They learned to communicate with the machines on the floor through the computer, utilizing feedback from the machines to improve their performance. This required a considerable educational effort on the part of the people.

Their success with the DNC system had a positive psychological effect on the engineers and programmers. The system was a solid professional achievement, with which they could (and did) identify.

The design and construction of the spindle/cuff cell also was a major undertaking for the engineers involved in the project, and they were understandably proud of their accomplishment. The team of engineers had designed a complicated production system, and they felt they had done it better than outside suppliers. The team had also managed the project, utilizing their suppliers for detailed design and construction of the equipment—all according to Sikorsky requirements. Thus, they had succeeded both in design of the system and management of the project. During our interviews the engineers and programmers showed strong feelings of identity with the company, linking its success to their own personal achievements.

Impacts on Operators

Involvement in selection and design. Both management and operators reported that operator input to the planning process was invaluable, and that it was a critical factor in the final decision on equipment purchases. Management at all levels—foremen, manufacturing engineers, programmers, and the Vice-President—expressed a genuine respect for the critical judgment of operators. They reported that operators were able to evaluate the capabilities and limitations of the equipment thoroughly, providing a level of evaluation beyond that of even the manufacturing engineers.

The result of this greater participation by workers was an apparent increase in managers' respect for operators. The informal skills and knowledge of the operators were evident during the machine selection process, often becoming formalized in written and oral evaluations and recommendations of equipment. Operators reported feeling more responsible for the successful operation of the machines, and greater involvement and pride in the production process.

Operators were also involved in evaluating initial plans for facilities and process design. In the original plan for the spindle/cuff cell, for example, cutting the threads on the spindle was done early in the machining process. In reviewing the machine layout and the process, operators noted that the threads were critical, that they would be difficult to protect as the part went through the subsequent machining steps. They suggested that the threading operation be moved to the end of the machining cycle, thereby reducing the chances of damaging the threads.

In another instance, an operator noted that the cutoff saw, which operated on an intermittent basis, was designed as an independent workstation, requiring an operator to run the saw. He suggested to the manufacturing engineers that they move the saw next to his machine. The radial drill he operated had long cycle times, so he could run the cutoff saw during the drilling cycle. These were just two of a number of suggestions operators made during design review that resulted in more efficient production.

Suggestions for postimplementation process improvements. The suggestion system appeared to be fairly well utilized and provided substantive recognition of operator contributions. Operators made suggestions for a number of changes to improve both work tasks for operators and quality of the products. For example, one operator suggested extending the overhead crane 4 or 5 inches to make it easier to move materials and perform his job. Another operator was dissatisfied with the time it took to tighten parts in fixtures. He suggested that hydraulic fixtures would be much faster and safer. This change dramatically decreased setup time and virtually guaranteed that parts could not be pulled out of the fixture during machining, increasing the safety of the operator.

The Vice-President's requirement that all operator suggestions have a formal response increased the attention engineers paid to the suggestions. His central staff began to follow up written suggestions by talking to the operator who made the suggestion. This increased the staff's understanding of the problems and the suggestions. By talking to operators, they discovered that many of the suggestions that had not seemed feasible in writing were, in fact, quite feasible and important. Written suggestions often were poorly communicated because of the difficulty of accurately describing ideas in writing and because of operators' lack of writing skills. This had previously led to a high rejection rate. The requirement for follow-up led to more careful investigation of suggestions and a dramatically increased acceptance rate for worker ideas.

The vitality of the suggestion system appeared to be related to several factors, and to vary over time. Overall, it was reported that the suggestion system became more responsive to operators (e.g., faster responses, more encouragement, better payout of awards) because of the emphasis on this system by the Vice-President of Machining and Transmission Manufacturing. However, the involvement of operators in the suggestion system also appeared to vary with the amount of publicity about the program. At the time of the study, there appeared to be little public recognition of suggestions. They were not reported in the company newsletter, and award checks were simply mailed out quarterly. One long-time operator on the third shift was only vaguely aware of the suggestion system, saying he had been frustrated by his current foreman in his attempts to recommend changes. Some of the managers reported a natural ebb and flow in the importance given the suggestion system. It would receive attention during the initial phases of a project, then decline in emphasis and attention until dissatisfaction built up. Eventually the problems of the suggestion system would receive renewed attention, and it would be revitalized.

Safety. The Vice-President's emphasis on safety, particularly eyeguarding, increased awareness of this safety policy. Most, if not all, employees—operators, supervisors, and management—were aware of the Vice-President's emphasis on this policy. At the time of this study the majority of the operators were complying with the glasses regulation, but a few were not. Subsequent to the study the company made eye protection mandatory in all manufacturing areas except a few aisles connecting office areas.

The nature of work in machining exposes workers to a significant level of potential risk of injury. Only one significant injury, however—the loss of a finger (replaced by microsurgery)—was reported to have occurred in the shop in the past several years. The most common minor injuries were from heavy lifting and from metal chips, which caused minor cuts and slivers.

Comfort. The layout of the machines and machining cells appeared to be convenient and to take operator needs into account. However, since these were mostly standard machines, worker involvement in selection was not always effective in ensuring operator comfort in designs. Overall evaluation of the machine required trade-offs among different factors.

One example, and a notable exception to the design of most of the machines, was the design of the Mazak turning machine. This was the newest piece of equipment in the area, purchased in 1988, but not completely accepted by Sikorsky at the time of our study. Its controls were located in a semienclosed stairwell of two steps next to the door into the machining work space, so the operator could have access to the controls while being close to the process, the tools, and the parts. The controls and screen, however, were positioned opposite the door, at the top of the stairwell, so the operator had to turn around on the step to go from the controls to watching the process. There was room for only one person to stand on the top step, which made it difficult for two operators or an operator and an NC programmer to consult while making changes to the program. The position of the controls did not provide for comfortable machine operation and posed a potential safety hazard by requiring the operator to be perched on the stairs in order to have access to the controls and screen.

Until Sikorsky accepted the machine, this arrangement was to continue, but then a platform was to be constructed that wrapped around two sides of the machine and covered over the stairwell. Workers had been involved in the selection of this machine, but because the design was established by the manufacturer they had no influence on this aspect of its design.

Control and modification of the process. In other locations in the shop where there were operator-programmable machines, it was found that operator programming of less complex parts was much faster than having the work done by the programming department. An operator-written program was usually completed within one day while the same program would take five days if it were handled by the programming department. Operators also commented that when the programming was done at the machine, whether by a programmer or operator, it was completed with fewer "bugs." The person doing the programming at the machine had the advantage of being able to view the actual tool path and machine in operation as it was being programmed.

We interviewed three operators who were programming their own machines. We asked how they would feel if they were assigned to machines that could not be programmed by them. None said they would like it. One said, "I want to do my own programs. It takes less time. Engineering hasn't operated the machines."

NC programmers were actively involved in teaching operators about programming and editing. The programmers generally viewed increasing operator-programming skills as a benefit, because it relieved them of the more mundane programming tasks. Yet, some managers expressed concern that programmers were not as forthcoming as they could be, and that they limited the amount of programming instruction given to operators. The programmers considered the operators as having limited ability to do programming, and some of the foremen agreed. Other foremen and managers thought there could be increases in the amount of programming given to operators.

The manufacturing engineers reported that the size of the programming staff had been reduced 50 percent in the last five years. At least part of this reduction was attributed to the increase in operator programming.

In general, the position of the foremen and supervisors seemed to be that operators ideally should have no reason to alter programs. It was felt that the programming department should have complete responsibility for ensuring that programs worked and for making any necessary modifications.

The scheduling system of the spindle/cuff cell had two contradictory impacts on worker knowledge and control of the machining process. Traditionally, machining jobs were piled around the operator's station and, within some limits, left to the operator to schedule. This gave the operator some flexibility in scheduling jobs as he or she saw fit, but, from the foreman's perspective, it also allowed some operators the opportunity to take the easier jobs from the pile. The automated scheduling system put limits on operator discretion. The scheduler, in conjunction with the foreman, established the order for jobs to be queued and delivered by the ASRS. The operator could take a job out of queue for certain reasons, but was limited in flexibility and was more directly accountable for doing so.

Although there was some negative reaction to this loss of discretion, there was also a positive response to the added information the scheduling system provided the operator. The operator could pull up his or her queue on the screen to see what jobs he or she would be working on. One operator noted that a benefit of this was to provide a longer-range view of the work to be done and to be able to anticipate jobs that might have special requirements (e.g., noting that there would be an especially messy job coming up and thus plan to wear overalls that day). The scheduling system also increased the speed with which parts moved through the machining operations. The operators reported liking this because they knew the jobs they were working on would be moved to the next operation in an orderly fashion instead of lying around for long periods of time.

Learning and skills. An unintended consequence of the design of the spindle/cuff machining cells was an increase in operators' interest in learning new skills and in cross-training. In the traditional machine shop, workers "protected" their skills. Because machines were segregated by type, operators seldom had the opportunity to focus on more than one or two types. In one instance, for example, only one operator knew how to operate a particular machine efficiently, and he refused to train other workers on its operation. He died unexpectedly, and there was a significant period of downtime and inefficiency while a new worker learned how to operate the machine.

In contrast, workers in the machining cells taught each other how to operate the different machines and expressed greater motivation to learn new skills. The design of the machining cells grouped together a number of different machine types and put operators with different skills in direct contact with one another. This led to workers wanting to learn how to operate the various machines. Operators began to cross-train each other with the encouragement of supervision.

To increase their understanding of the machines and machining operations, some workers began taking computer and math classes at the local junior college. These operators would then expand their skills, using their new knowledge. One operator explained that the operators who had been taking classes would explain to other operators that they were determining machining operations by "trigging it out," referring to their use of trigonometry. In this atmosphere, operators encouraged each other and felt challenged to go to school.

Operators in the machining cells reported they had a broader view of the entire machining process. They could follow the progression of a part from one machining operation to another. The result was that operators could see the significance of each operation and the interdependence of the different steps. This broader view of the production process and greater knowledge enabled operators to make more, and better, suggestions for improving the process.

The company often recognized the new knowledge and skills with promotions. The clearly recognized opportunities were for advancement to new positions, particularly out of machine operator positions to supervisory jobs, or, with additional schooling, into programming or engineering departments. The company, however, had not yet settled the question of recognition in machine operator pay grades, of new capabilities such as multiskilling, multiple machine operation, or increased knowledge.

In summary, the design of the cells brought workers into closer contact with each other and provided a broader view of the machining process. The unanticipated results were the formation of informal work groups, informal cross-training instead of "protecting" knowledge, more enthusiastic and effective suggestions, and increased motivation to attend school and learn new skills and knowledge.

Impact on Product

The effect of the new cell on the spindle and cuff parts was felt in several ways. Of major importance, both in cost and in performance, was the greatly improved yield of acceptable parts. The scrap rate (loss of material for all reasons) for these expensive parts was reduced by almost two orders of magnitude, a remarkable achievement. Product throughput time was reduced by 50 percent or more, and capacity was increased at least threefold. This enabled Sikorsky to bring back to the shop work that had been farmed out to vendors at premium prices because of lack of capacity.

This expanded capacity gave Sikorsky greater control over parts that were critical to helicopter performance. It also provided Sikorsky with two competitive weapons—lower product costs and faster response to new orders. Not only could higher profits be made on units already under contract, but more attractive terms could be offered for subsequent helicopter procurements.

7

THE TIMKEN COMPANY FAIRCREST STEEL PLANT

THE COMPANY

The Timken Company was founded as the Roller Bearing Axle Company in 1899 as a manufacturer of tapered roller bearings. Because of the shortage of steel caused by World War I, it began producing its own steel to supply its bearing plants and, at the time of our case study, was a major source of high-quality, low-cost alloy steel made specifically to customer order.*

Timken's annual sales in 1987 were reported as just over $1.2 billion and total employment as 16,721. Timken was one of the top three companies in terms of sales in the antifriction roller bearing market (SIC 3562). With the productivity gains being made at its new Faircrest Steel Plant and in other Timken steel plants, the company was gaining in market share in cold finishing of steel shapes (SIC 3316) and steel pipes and tubes (SIC 3317). Timken appeared to be rapidly establishing itself as the low-cost, high-quality supplier in the alloy steel market. Timken had been experiencing severe competition from a large group of foreign bearing companies in past years, but had recently obtained favorable decisions from the U.S. government in several antidumping cases. Of its 1987 sales, 61 percent involved bearings and 39 percent, steel.

A five-year summary of net sales, net income, and a three-year summary of cash flow from operations is presented in Table 7–1. Both net income and cash flow from operations reflect the competitiveness of the market and the commitment and extensive involvement of the company with the Faircrest Steel Plant (FSP) during

*Dates of on-site visits: December 12, 13, and 15, 1988.

Table 7–1 Timken Net Sales, Net Income, Cash Flow from Operations, 1983–1987 ($ millions)

	1983	1984	1985	1986	1987
Sales	937	1150	1091	1058	1230
Net income	1	46	–7	–83	10
Operating cash flow			28	11	103

Sources: The Timken Company Annual Report, 1987
Standard and Poor's

this period. The 1987 figures, however, indicate the positive contribution being made by the FSP, which began to reach, and in some instances exceed, expected production levels in that year.

Competitive Environment

The markets in both bearings and alloy steel were very competitive, although each was generally dominated by about four or five companies. Timken's customers were essentially all industrial firms.

In its 1987 Annual Report, Timken stated the company's mission as follows:

> We are an independent organization with a leadership position in quality anti-friction bearing and alloy steel products. To maximize shareholder value and sustain our competitive position, we will capitalize on the competitive relationships between our businesses and emphasize the application of technology to products and processes combined with unmatched customer service.

The Annual Report also stated:

> Since we are both a very demanding user of steel for the manufacture of tapered roller bearings and a producer of high quality steels with the metallurgical expertise to advance steel product performance, a very real synergy exists. Our organizational structure brings each customer into closer contact with more of our people than ever before . . . We reduced our costs and lowered our breakeven point while retaining our . . . traditional strengths of research and development, design assistance, . . . technical support, and . . . managerial talent.

Location

Timken's general offices were in Canton, Ohio, which was also the location of one of their bearing plants and the Faircrest Steel Plant, the focal point of this case study. Timken had 79 plants and offices in 11 countries on five continents, including a steel products plant in Latrobe, Pennsylvania. To avoid potential overcapacity, Timken was phasing out a standard bearing plant in Columbus, Ohio.

Management and Organization

Timken was originally a family-owned company, and several family members were still involved. Three were on the Board of Directors, which appeared to be both innovative and actively involved in the strategic affairs of the company. The company organization and management style had been quite traditional with a hierarchical line and staff structure and decision making based on "bottom-line" performance and return on investment.

The line organization was divided into the Steel Business and the Bearing Business, each headed by an Executive Vice-President who reported directly to the President. Each Business Unit was responsible for its own planning, manufacturing, marketing, quality assurance, and technology support.

The company staff was divided into four groups called Corporate Centers. They were: Strategic Management, Financial and Legal, Technology Center, and Personnel and Logistics. Each was headed by a Vice-President who reported directly to the President.

The Technology Center consisted of four groups: Research and Development, Applied Development, Computer-Integrated Business, and Management Information Systems. The two development groups were engineering oriented. MIS was a computer science development and support group. Computer-Integrated Business was involved with the development and integration of computer control, information, and decision-support systems.

Faircrest Steel Plant (FSP)

In 1978, the Timken Board of Directors, under competitive pressure from both domestic and foreign steel producers, and encouraged by an expected boom in oil and natural gas production and the accompanying requirements for steel, called for the updating of the company's steelmaking technology.

Studies were made of the technologies in use and under development in the major steelmaking countries throughout the world. These were followed in the early 1980s by visits by teams of Timken engineers to a number of plants in several countries. Their purpose was to examine the technology firsthand and to determine its effectiveness and reliability. The technology considered to be the best for each phase of the steelmaking process was identified, and from 1983 to 1985, Timken engineers and technologists were sent for extended visits to the plants where each selected technology was being researched, developed, and/or used, to learn all they could about the technology, to participate in its final design and development, and to gain experience in how to use and maintain it effectively. For example, one of the members of the Computer-Integrated Business Group spent almost two years at Krupp Steel Corporation in West Germany learning about and working with what at that time was regarded as the latest and most effective electric furnace technology in the world. A similar extended visit was made to a company in Toledo, Ohio, to assist in the development of the latest and best technology in ladle refining. Similar activities were performed relative to scrap management and handling, ingot pouring, soaking pit design and operation, rolling mill design, and several of the 14 other steelmaking technologies represented in the plant.

Through this investigation, Timken had identified the best available technology for each step in the steelmaking process for the set of products it would be producing. The results of this study led the Board of Directors to decide to make a clean sweep and invest in a new plant, rather than simply update existing plants. This move would also provide the opportunity to develop simultaneously a new, more effective organizational structure, shop-floor and management culture, job design, computer aids, information systems, and plant and pulpit layouts. (Control rooms were called "pulpits.") Likewise, a work force could be selected using criteria that would be more readily compatible with the new technology, the anticipated computer interfaces, and any new, creative job designs and work organizations. In this way, old, inhibiting habits and attitudes could be avoided. In 1982 a 10-year no-strike union labor agreement had been signed that allowed the company to select from any Timken plant the people who best "fit" the types of jobs envisioned for such a plant. Soon thereafter the decision to build was made and $500 million was designated for the project, an outlay close to two-thirds of the net worth of the company at that time.

Sites were evaluated, and early in 1982 a 450-acre site was selected on Faircrest Street in Canton. Engineering and construction began soon thereafter. Almost simultaneously, however, the bottom dropped out of the oil and gas markets and the railroad industry became essentially "flat," making it very unlikely that the planned Faircrest rating of 500,000 tons annually could be realized. The Timken management, however, made the very difficult decision not to cancel the project and not to reduce the $500 million budget. Their only admonition to the plant design team was to be careful in their planning and to justify in their own minds whatever expenditures they felt necessary. These circumstances caused some design changes to satisfy changes in market expectations away from a steady flow of 160-ton orders for a small variety of products to many much smaller orders for a wide variety of products.

Personnel mobilization was initiated in early 1985, and start-up began on August 5, 1985. Full four-crew operation was attained in April, 1986. Table 7–2 provides information about the plant and its construction.

Table 7–2 Faircrest Steel Plant Data

Facts About the Plant:	Facts About Construction:
$500 million investment	450-acre site
1/2 million annual tons output	20.6 acres under roof
550 employees (Phase 1)	3,230,000 yards of earth moved
Team concept	25,000 tons of structural steel
127,000 training hours	70,000 yards of concrete
560,000 computer development hours	12 miles of railroads
Daily electrical use = 50,000 homes per day	750,000 engineering hours
Daily natural gas use = 62 homes per year	2,650,000 construction hours
Annual scrap use = 440,000 cars	

Source: Faircrest Steel Plant fact sheet

The Mission and Objectives of FSP were stated in a company memorandum dated November 22, 1983:

Mission

The overall mission of the Faircrest Steel Plant is to generate income potential for the Company by being the lowest cost producer of high quality, low alloy steel in the world for our particular market. We will achieve the mission when we successfully satisfy the needs of the following interested parties on a continuing basis:

Our customers—who expect quality products, on time and at competitive prices.

Our shareholders—who expect us to maximize the return on their investment.

Our counterparts in other parts of the Company—who expect us to give and receive services and products so as to most efficiently and effectively use Company resources.

Our immediate employees—who expect secure employment and satisfying work.

Objectives

The extent to which we have sustained accomplishment of our mission in the long run will be determined by the quality of our personnel and by the quality of our technology, and our commitment to long-term improvement. We therefore have three long-term objectives:

1. Personnel Development
   To have people who are able and motivated to accomplish the Plant mission.
2. Technical Development, Implementation, and Independence
   To keep the Plant ahead of the state-of-the-art by pursuing corporate development of steel making technology, by rapidly and effectively implementing improved technology and by continually improving existing methods and technology. (The capability of providing increasingly higher qualities of product will result from this objective.)
3. Continual Improvement
   To continually improve the performance of the Plant in all areas— quality, timeliness, cost, investment utilization, and work environment.

The mission and goals set forth for the FSP had a significant impact on both the facility design and the organizational structure of the plant. Later sections of this report provide details on the design process used to implement these objectives. Unless specified otherwise, all material in the remainder of this report refers to the Faircrest Steel Plant.

THE PRODUCT

Timken product sales literature indicated that, "In 1916, a wartime steel shortage prompted The Timken Company to begin producing its own steel." During the late 1970s a consensus began to form in the company that additional sources of high-quality, low-alloy steel were needed. In mid-1981, Timken management announced plans to build the $500 million Faircrest Steel Plant.

Timken's Faircrest Steel Plant used scrap steel as its major raw material in producing high-quality, low-alloy steel in various forms, including conditioned billets, unconditioned billets, blooms, and ingots.

A conditioned billet was a round bar or a round-cornered square bar of steel that was given special grinding and finishing treatments after it cooled. An unconditioned billet was the same item without special treatments. The round billets from FSP ranged in diameter from 6 through 11 inches, and the square billets had sides between 6 and 12 inches. The billet ranged from 10 to 35 feet long, according to exact customer specification. Blooms were square steel bars that had loose tolerances for length and cross-sectional size. Ingots from FSP were square with sides of 24 or 28 inches and had no treatments to finish their surfaces. Any of these items could be sold to customers. Billets and blooms were also shipped to other Timken facilities for use in manufacture of tapered roller bearings or steel tubing or bars.

In addition to controlling the size and finish of the steel, FSP adjusted metallurgical properties to produce up to 700 different steel alloys. At the time of this study, approximately 350 different steel alloys were being produced.

Faircrest and the other Timken steel plants operated in a product area that lacked clear definition. Stainless steels, tool steel, and superalloys all had recognizable specialty classes, but Timken's alloys fell somewhere between carbon steel and the highly alloyed grades. (McManus, 1987).

The production process had several features to assure a high-quality, low-cost steel. Costs were kept low by a combination of factors including:

- High volume
- Simple once-through process without any provision for rework of defective items
- Extensive computer-based assistance for operators
- No parallel processing paths that required duplicate equipment
- Minimal work-in-process inventory

The high volume of FSP was due to the size and speed of the equipment.The electric arc furnace that melted the scrap held up to 160 tons of scrap and melted its contents to customer requirements in less than 2 hours. The furnace was relatively large and the two-hour process time was one-half to one-third of the common industry process times.

FSP included provisions for recycling small amounts of scrap that occurred when steel was cut or finished, but there were no dedicated rework areas. FSP management reported that the process rarely produced an entire batch that had to be scrapped. Recycled scrap from FSP was included as raw material for future "heats" of FSP's electric arc furnace.

Computer assistance for operators reduced cost by calculating the cost of operator actions and proposing other actions for the operator. The proposed alternatives were available to the operator within 10 seconds of the time an inquiry was made. There were also computer-based communication systems that informed operators of production requirements for current and future product at their station, described the status of product at all stations, described what previous operating stations did to the product currently at their station, and described the current metallurgical properties of the product at their station. This information facilitated operator decision making, allowed the operator to know the cost of alternative courses of action, and eliminated the supervisor from the process.

Because there were no parallel production processes, the production sequence for every item made by FSP was the same. This minimized the complexity and cost of planning the sequence of the various batches of steel through the plant.

The plant had almost no work-in-process inventory. From the time that enough scrap steel was collected for the next "heat," until the time that hot, unconditioned billets left the rolling mill, the only time that product was not in motion was when it was being treated by a stationary operation. Some cooled billets may have waited in a cooling yard before entry to the billet-conditioning shop, but there appeared to be less than one week's production in the cooling yard.

FSP's steelmaking process included unique features intended to assure high quality. The *ladle-refining* process was similar to the chemistry titration process taught to high school students, except that the refiner operated on tons of molten steel and added up to eight different alloys to the ladle. This process was done under vacuum to remove gaseous impurities from the molten steel. Ladle refining was guided by rapid feedback from a metallurgical assay lab. Ladle contents were stirred by injecting argon gas into the melted steel. If it were not for its high-quality requirements, FSP might not have had a ladle refiner or might have used a less sophisticated refiner.

Molds were used to form ingots as the molten steel cooled. The required quality of the steel dictated that special care be used when pouring molten steel from the ladle into the molds. FSP used a *bottom-pouring* technique that filled molds through the bottom to improve surface quality of the ingot.

The requirement for a high quality product influenced the designers' choice of the plant's *once-through process.* Their intent was that each part of the process be done right the first time.

The metallurgical quality of FSP's high-quality, low-alloy steels was dependent on the mix of alloys in the steel and the amount of hydrogen entrapped in the steel. Alloy mix was influenced by the nature of the scrap loaded into the electric arc furnace, the metallurgy of the residual metal in the bottom of the furnace when a fresh batch of scrap was added, and the refining process. Hydrogen gas that might have become trapped in the steel as it cooled was removed in the ladle by vacuum and argon gas stirring, and the bottom-pouring technique reduced chances that gas could become trapped while steel was poured into molds.

Surface quality was also important for the product. The bottom-pouring technique described above and another technique known as scarfing helped improve surface quality. Scarfing was done with a stationary ring of gas torches in the rolling mill. The torches burned off a thin layer of the surface of the steel blooms as the blooms moved through the rolling mill.

The billet-conditioning shop had an automated nondestructive test to locate hairline cracks on the billet's surface. The automated equipment marked the cracks with paint, and personnel in the billet-conditioning shop worked on the surface with high-speed grinding tools to remove the hairline cracks.

The plant's high volume affected the market for raw material. FSP used so much scrap metal that their purchases could have had a major influence on the local price of scrap. Scrap purchases were adjusted to control this problem. The intent was to meet the company's needs without demanding so much from suppliers at any one time that prices rose.

THE PROCESS

Steelmaking at FSP included the following operations:

- scrap loading
- scrap melting
- ladle refining
- bottom pouring
- ingot soaking (heating)
- bloom forming
- bloom scarfing
- bloom shearing
- billet forming
- billet ID marking and cooling
- billet conditioning

Each of these operations is described in the next section. All of the computer-assisted operations had a computer-generated display that showed the status of all heats of steel in the melt shop and rolling mill. Following the description of the steelmaking steps is a description of operator responsibilities at FSP. Figure 7–1 provides a schematic layout of the plant.

Steelmaking Operations

Scrap loading. The scrap-loading operator occupied an equipment control room placed on the outside wall of the melt shop above a four-track railyard. He had control of a crane that carried an electromagnet to move scrap from rail hopper cars to a scrap bucket. The operator could control movement of the railcars and the scrap bucket. The scrap-loading station was staffed by two operators. One operator con-

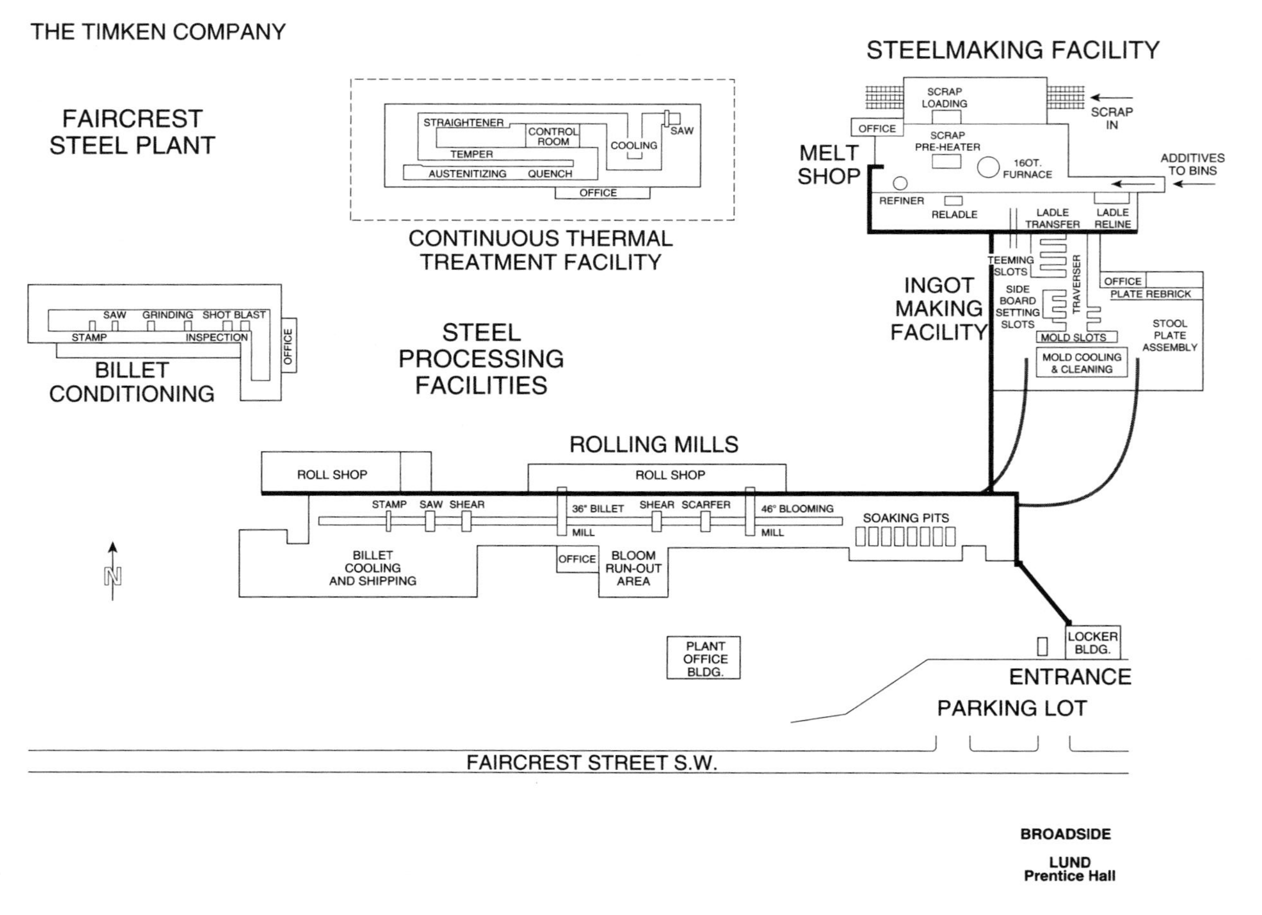

Figure 7–1 Faircrest Steel Plant layout

trolled the equipment and the other removed data sheets describing the contents of the railcars from each car and brought the sheets to the scrap-loading pulpit. At this time, the quality of the scrap in the cars was checked.

Computer support for the scrap-loading pulpit included systems that selected railcars for the next four heats or batches of steel, told the operator the cost and technical feasibility of a selected mixture of scrap (scrap-optimizing system), and told the operator the contents of all rail cars under his control. The operator loaded the scrap bucket from the scrap-loading plan prepared by the scrap-optimizing system. In addition the system gave him the cost and feasibility of changes he might make to the recommended plan.

Scrap melting. The furnace operator controlled scrap melting.His tasks included preheating scrap with waste heat from the furnace, loading the furnace, melting furnace contents to a desired metallurgical state, and controlling delivery of the alloy additives placed in ladles that received molten steel from the furnace. There were two furnace operators in the furnace pulpit and three attendants who performed tasks at the furnace.

The furnace operator brought scrap steel to the furnace by controlling the movement of the scrap bucket containing up to 160 tons of scrap. The operator preheated the scrap while it was still in the scrap bucket. The scrap bucket was then emptied into the furnace and returned to the scrap-loading area to be refilled.

Computer support for the furnace pulpit included systems to monitor the condition of the furnace, to monitor utility usage by the melting operations, to control material handling, and to communicate with the metallurgical assay lab. These systems helped the furnace operator determine when the scrap had been properly melted to obtain useful samples of furnace contents for metallurgical assay, and to empty the furnace contents into the ladle for the next step. The operator had to properly melt the scrap while keeping the operation on schedule, minimizing wear on the furnace, and minimizing power usage.

Ladle refining. Ladle refining required one operator in the ladle-refiner pulpit and another who connected utilities and sensors to the ladle. Molten steel was assayed and mixed with alloys to give the product precise metallurgical properties. Hydrogen and other gaseous impurities were removed from the molten mixture using vacuum and argon stirring.

The refiner operator had computer assistance for material handling, for control of the refining process, and for communication with the metallurgical assay lab. A computer system informed the operator of a least-cost mix of alloy additives that would change the ladle contents to the desired metallurgical state. The operator could use the mix planned by the computer or have the system give the cost and quality of other mixtures proposed by the operator.

Bottom pouring. Timken literature described the bottom-pouring operation as follows, "Liquid steel is teemed (poured) into a circular cluster of 6 or 8 ingot molds. The molds are connected at their bottoms by refractory pipes to a central trumpet. The steel is teemed into the central trumpet filling all molds at the same time from the bottom up." Four clusters were poured from one heat of steel.

Ingot "soaking". This process was controlled by an overhead crane operator who removed ingots from molds, placed them in a refractory-lined pit, and monitored the ingot temperature. The pit used computer-controlled gas heat to bring the ingots to a uniform specified temperature for proper performance in the rolling mill.

Bloom forming. After the overhead crane moved the ingot from the soaking pit to the rolling mill, the 24-to-28-inch square ingot was rolled into an 8-to-15-inch square "bloom". The blooming-mill pulpit had two operators and extensive computer assistance. The computer control system controlled the direction, speed, and alignment of the product and the operation of the bloom mill. The rollers moving the blooms could not withstand long exposure to the hot blooms so the operator also controlled the motorized rollers to constantly keep the product in motion.

Bloom scarfing. Timken described scarfing as follows, "As the bloom passes, gas torches burn and blow away a thin layer on all four sides of the product. Scarfing removes surface scale and provides a clean surface for further processing."

Bloom shearing. The operator in the blooming-mill pulpit controlled the bloom shear and the alignment of the product under the shear to remove unwanted material from both ends of the product and to cut the bloom to required length. After shearing, some blooms were removed from the mill and sold or shipped to other Timken plants. Other blooms continued processing in the billet mill.

Billet forming. Special rollers pressed the blooms into round or square billets. The billets were cut into 10-to-35-foot lengths.

Billet marking and cooling. Computer-controlled pins placed a dot matrix ID on the end of each piece, then the product was cooled. Product to be control-cooled was covered with an insulated hood and moved to an outdoor cooling yard. Other product was air-cooled first on a cooling bed at the end of the rolling mill and later on pallets in the outdoor cooling yard.

Billet conditioning. This shop performed special inspection, surface finishing, grinding, and cutting operations on the cooled billets.

Work Environment for Operators

During the early weeks of FSP design, Timken decided to create a new "culture" for workers at FSP. One of the changes for the operators was that they were expected to act as a team with minimal supervision. There were no steps of the process that routinely required work by a supervisor. Operators were expected to validate a computer-generated recommendation whenever possible. The computer system supported this culture by giving operators predictions of other decisions and feedback about results. The predictions and results were provided at operator request and were available to the operator within 10 seconds. Another new expectation of FSP operators was that they use the computer system for first-line diagnosis of equipment problems. This reduced downtime and maintenance costs.

The team concept had a major impact on operators and maintainers at FSP. Operators and maintainers performed multiple jobs. All hourly workers at FSP were expected to contribute to ongoing improvement at FSP and demonstrated considerable success in originating and applying improvements.

The once-through process and plantwide status monitoring system combined to show the results of each operator's decisions about the product to all other operators in the plant. Since all of an operator's peers had immediate information about an operator's performance, there was a strong incentive for high performance. Another day-to-day effort to assure high performance was the FSP production schedule, which gave operators the planned sequence of future heats for the next several days' production. The simple flow of the process reduced the complication of the production schedule and allowed operators to focus on quality aspects of the process.

Anticipated Process Improvements

Three key items of the process Timken had expected to improve were plant throughput, process control, and operator involvement. Timken personnel were very excited that FSP was producing more steel than expected. FSP came close to completing 14 heats in a 24-hour period. Process control improvements had arisen because of the process developments and because of an ongoing effort to provide better information to the operator. FSP management was seeking better methods of employee involvement for planning process improvements. Situations where attempts at employee involvement yielded little employee contribution prompted searches for new methods for employee involvement, rather than criticism of the employee involvement concept.

Computer-Based Support of Process

In addition to assisting operators, FSP's Hierarchical Computer Control System (HCCS) was designed to support production planning, material tracking, quality assurance, maintenance, and inventory control. Operations support and other functions were integrated into a system that allowed access to almost all of the system's information from terminals in many locations in the plant. Dravo, the firm that managed the design and construction project for FSP, described the computer system as allowing information to be shared between the operator at the plant floor level and the executive at the corporate level. The hierarchical levels of control were as follows:

level 0. Sensors and actuators.

level 1. Programmable logic controllers for machine control, data acquisition, and communications. This level included Gould Modicon PLC's connected to the level 0 equipment and included distributed Digital Equipment Corporation (DEC) PDP 11/23 microprocessors to provide a gateway between level 1 and level 2.

level 2. PDP 11/44 computers which ran the CRISP software package from ANATEC. This level provided supervisory control for the process, disseminated process information from level 1 and provided an interface for the operator. The level 2 computers were connected to one or more level 1 computers with a local area network (LAN). Another LAN connected all level 2 computers with each other and with levels 3 and 4.

level 3. DEC PDP 11/44 computers to provide shop management support for an area at FSP. For example, there was a level 3 system which supported the melt shop (scrap loading, furnace, ladle refiner).

level 4. DEC VAX 780 computers to perform plant management functions, such as reporting historical data.

level 5. The steel division's IBM mainframe computers, remote from FSP. These performed traditional corporate functions, such as order entry, billing, and payroll.

level 6. Mainframe computers for company-wide functions.

Computer support gave production personnel such a high level of control that their product met specifications more consistently than product from plants with less process control. The computer network and information sharing promoted ongoing process improvement by giving many people the information needed to identify what could be improved.

Computer implementation of process control also simplified some improvements by allowing them to be done by rewriting the software rather than by changing the hardware. FSP managers used this feature of their process control system carefully. Their intent was to avoid using software changes in cases when poor hardware performance was a problem. The managers did not hesitate to make hardware changes when the hardware was at fault.

DESIGN PROCESS

Background

The construction of the Faircrest plant was the result of a high-risk decision made by Timken. The financial viability of the company depended on the project's success. As one company representative put it, "We were literally 'betting the farm' that we could do this." The resulting facility successfully used process computer systems to integrate the best of the world's steelmaking technology.

The Timken organization took a very different approach to the design of the Faircrest plant. The actual phases and dates of the project are summarized in Timken's milestone chart (Figure 7–2). This chart shows the sequence of activities involved in the project, but does not convey the full significance of the change in approach to project design. The Faircrest organization was a radical departure from Timken's traditional top-down management style. Early strategic decisions and top management's willingness to remain committed to those decisions led to the development of an extremely dedicated design and management team. Two of these decisions, in particular, had a major impact on the design and development of this plant. They were made during the "development of technology" or feasibility study phase from 1978 to 1981.

The first of these decisions was that the technology was to be the best in the world and the most up-to-date version of a proven technology. The Timken people were experts in steelmaking. To use this expertise they wanted equipment that would allow them to reach the highest product quality and most accurate chemistry possible on a consistent basis. They did not want to experiment with unfamiliar processes. There was to be no redundancy or backup equipment built into the plan. There was to be only one furnace, one overhead crane, and so forth. All equipment, therefore, had to be both reliable and easy to maintain in order to minimize equipment downtime. This was also to be a once-through line so there would be no provisions for rework.

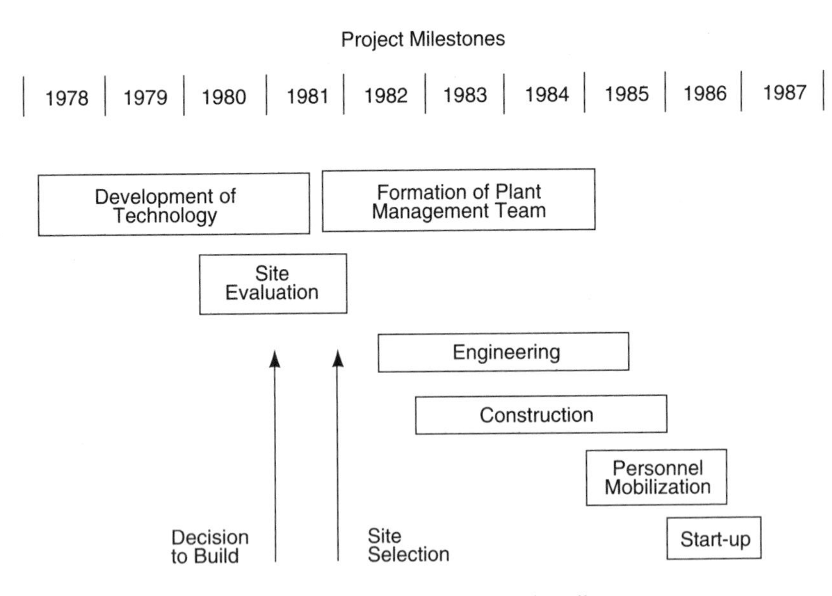

Figure 7–2 Faircrest Steel Plant—Major milestones

In order to get the superior equipment to accomplish this, Timken was willing to go worldwide to purchase the technology for the individual components of the steelmaking process. It is important to recognize that, although there was no radical change in the steelmaking process itself, the bringing together of these state-of-the-art components into a computer-integrated factory would be creating a high-technology environment. This would mean many changes for those who would have to operate the facility.

The second decision had to do with the management philosophy for operating the plant. During this early planning phase it became apparent to senior management that this plant was to become a very different sort of steel plant. It occurred to them that this might also be an opportunity to create a work environment that would be a major departure from the adversarial labor-management relationships so common in the steelmaking industry.

There was a desire to make a change in management philosophy and leave old habits behind. This decision brought the human resource planning function into the design process at a very early point. This was to have an impact not only on the way in which the design team functioned but also on personnel selection, compensation, training, management philosophy, supervisory practice, and control of the plant. The site selection was made during the latter stages of the feasibility study. To change the work environment, major adjustments by both management and the union would have to take place. For management, this would mean changes in supervisory practices and style. For the union, it would mean concessions on wage rates, number of job classifications and work rules. Site selection, which occurred during the latter stages of the feasibility study, was dependent upon union acceptance of the changes. The Faircrest location could be a feasible site only if the concessions were made. At first the union rejected the proposal, but later reversed the decision in order to preserve jobs in the Canton area.

Design Approach

The latter part of 1981 represented, for most people, the beginning of the Faircrest project. By that time the funds had been committed to the project and the more visible signs of the project began to appear. The Steel Project Team of engineers and human resource personnel was formed. Maintenance personnel and operators were not part of the project team. Members of the team were chosen for their assignments on the basis of their ability to learn quickly and their ability to work with others. Selection was made only after interviews, personal recommendations, and a four-hour battery of tests.

A new human resource philosophy for operating the plant had to be developed. To underscore the changes that were to be made, the team members were transferred from existing jobs and were fully dedicated to the project. They were brought together to work in one location to maintain close proximity to each other. Their assignment was twofold: (1) to design the plant and integrate the systems, and (2) to develop the operating philosophy for the plant. This created two distinct sets of tasks for the team: design engineering and human resource development. Members of the team described these activities as a process of having a common origin in conception, of separating and running in parallel paths for about 18 months, and then of recombining at the point of personnel mobilization.

Some members of the Steel Project Team were those who had been part of the original feasibility study. Others were chosen on the basis of psychological testing and compatibility with the organizational objectives of the project. For want of a better term, "matrix management" was the name given to the organization of this initial team of key people chosen for the Faircrest assignments. Those who were part of this group later suggested that "the gang" or the "marauding band" might be more accurately descriptive of the team. These terms seem to characterize the close working relationships and the freewheeling style that was associated with the project team. Although they were not explicitly told this, these engineers and steel experts who were designing the plant and operation understood that they were to become the first management team for the plant. This served to reinforce the commitment to the project and instill a sense of ownership in it. It also meant that these people would have to live with what they designed and would not have the option of walking away from any problems that their decisions created.

The new decision-making process represented a significant change for these people. Responsibility for major decisions was placed on the engineering and management teams. Decision making and prior approval were no longer the prerogative of senior management. The project had been justified as a whole before the funds were committed to it, so there was no requirement to cost justify the individual components. It was expected that people would base their judgments on what would be the best total value to Timken, as guided by steelmaking experience. People found themselves taking responsibility and making decisions of considerably greater magnitude than they ever had at any time in their careers. Initially this freedom was not always accepted. Often people would attempt to push decision making back up to senior management only to find it returned firmly to them. Eventually the people became comfortable with this and found it exhilarating. This excitement caught on with the project team and personal commitment grew rapidly.

The project team rapidly increased in size. To keep the growing team together as a group, several moves were made. Eventually the team moved to the site location and was housed in office trailers. Within the team there were smaller units organized around each of the eight different process areas. These units included both operations management and engineering personnel, and, in accord with the matrix concept, some individuals would be performing both functions. By whatever name it was called, the organizational terminology appears not to have been as significant as the group cohesion that developed by having a common location and a central focus.

At first Timken thought that it would be able to develop the factory automation systems for this facility with its own personnel. Even though there were 182 Timken people dedicated to the project at its peak, it became apparent that it could not be completed without adding a significant number of people. Rather than hire more people and then have to lay them off at the conclusion of the construction phase, a contractor, Dravo Construction Services, was engaged to help with the project. Dravo's resources, combined with Timken's, provided the Faircrest project with the 600,000 computer-development hours, 750,000 engineering hours, and almost 3 million construction hours required to design and build the plant.

Dravo normally operated as a turnkey contractor. It would make all decisions related to design and construction of a facility and turn over the completed facility to the client. At the Faircrest project, however, Timken reserved the final decision-making authority for itself. Dravo had to make an adjustment to this mode of operation. The Timken people had never worked with a contractor on a project of this magnitude, so it was a new experience for them also. Dravo was given much credit by Timken's management for its critical role in the design process. It was viewed as a demanding taskmaster, one who required strict adherence to timetables and development of specifications. If critical decisions were not made on time, Dravo would make them by default. Dravo would tend to choose the low-cost alternatives while Timken favored the preferred steelmaking technology. Therefore, in order to protect their decision-making initiative, Timken would have to see to it that decisions were made on time. As had happened within the Timken work groups, close working relationships also evolved among the Timken and Dravo people. The working meetings between these groups were described as often loud and strained. Two years after completion of the project, however, we heard not one word of criticism of Dravo from the Timken people. Many sources within Timken gave Dravo major credit for making the project happen. The consensus was that Timken knew how to make steel and Dravo knew how to manage projects.

Timken did much to establish excellent working relationships with its equipment suppliers as well. For each of the eight processing areas, Timken sent one of its employees to the supplier's site for up to two years of training. These Timken people worked as employees of the supplier, learning the technology as the equipment was being built. These people were then expected to document what they had learned in order to return home to teach the technology to other Timken people. Initially there was some reluctance on the part of the suppliers to accept these people into their organizations. There was some suspicion that they would be there to act as spies. This reluctance was quickly overcome and, here again, lasting friendships developed among those who participated in the project.

The individual process segments were designed by and for the operating groups who would run them. Eventually these individual components had to come together in the process stream at the Faircrest plant. The design strategy was: first, "get the process right" through mechanical and electrical design; then, get a neat package of equipment, controls, and computer; next, attend to the human interface; and finally, integrate the management information system (MIS) and recordkeeping. The computer systems were considered to be elaborate, but low-risk, systems. There were manual overrides for all equipment so that the processes could be operated without benefit of the computer systems.

While this description of the computer integration is consistent with the strategic decision to use only low-risk proven technology, it tends to understate the value of the computer systems for feedback and decision-support. For example, it would have been impossible to achieve in any other way the remarkable precision of the chemical composition of the individual melts that was achieved by linear programming using data from the furnace charge, the test specimens, and the ladle-refining process. It would also have been very difficult for the furnace operator to control the furnace without the in-process feedback on furnace conditions. The material handlers who prepared the furnace charges were able to optimize loads based on both the alloy content of the scrap and its general density. These were the production elements that gave the plant its competitive edge and the opportunity for continuous improvement of the process. The value of the computer controls was increasingly recognized as the facility design progressed. The budget for computer support grew from $3 or $4 million to $30 or $40 million but even at that it represented only a small portion of the total project cost. It was considered a minor expenditure to achieve such a significant impact on quality.

Human Factors Approach

During the same period that the engineering design was evolving, the parallel path of organizational design was being led by the training group. These activities were driven by the strategic decisions to change the organizational culture from a top-down managerial hierarchy characterized by adversarial relationships on the shop-floor to one that was broad and flat. It centered process ownership and control at the shop floor level. This meant significant changes for those who had a long experience in the steel industry. Early in the design process, at the same time that the plant management teams were being put together, staff functions for training, industrial engineering, and personnel were organized. There was clearly a concern about how human resources would be integrated into the process. Recognizing the need to change an organizational culture was one thing; making it happen took hard work. By mid-1983, after a series of meetings over a six-month period, the training team had prepared documents that outlined the mission, objectives, and operating philosophy of the Faircrest plant. These documents outlined a number of explicit human factors policies to be followed in the operation of the plant. Many of these also affected the way equipment and controls were designed. Using these documents as a pattern, each of the operating-area teams was expected to tailor similar documents specific to their requirements.

The area managers described how the change in human resource philosophy was carried out. The training group initiated Friday afternoon sessions for discussion on how the plant would operate. The expertise of the individual worker was to be recognized and respected. Those who were close to the process were to be given authority and resources to make changes as appropriate. The role of the supervisor was to change from one of being a "boss" to one that was more technical, concentrating more on data analysis and determining the causes of equipment failure. It clearly called for a person different from the traditional steel foreman. The people who attended these Friday meetings not only changed their perspective on how the plant would operate but also had an opportunity to practice their new roles. The training group provided role-modeling exercises to show how to handle common interpersonal situations. Videotape feedback was provided. Those who took part in these sessions reported that these new managerial techniques were hard to learn after years of using a combative style. They also found that this was not to be an abdication of managerial responsibility but rather a change in the manner of dealing with people. Gradually, a new culture clearly different from the traditional steel environment evolved.

Another clear break from the past was the reduction in the number of job classifications from 84 to 28. Operators would be expected to be flexible and willing to do what was necessary to make the process operate. The response, "It's not my job!" was not to be tolerated. People were to be given control of the processes and would be expected to do what was necessary to keep them running.

Operators received training to help prepare them for new roles in a new organizational culture. The technical training was also designed to foster group cohesion and close working relationships between engineering and operating groups. Provisions were made for technology transfer from the equipment supplier to the Timken plant by a learn-then-teach strategy. Those people who spent time at the vendor sites were expected to pass this information on to others by writing the process-documentation and training manuals. The engineers were also expected to take part in the training activities. At the start, the Faircrest training staff consisted of a manager and three training specialists, but at the peak of training activity this staff was increased to 10. The primary function of this group was to teach the technical experts how to write sound training manuals and plans. Because the training load was too heavy for even these 10 people, the training function became one of "train the trainer." Everyone, even those who were transferring from other Timken locations, had to be trained for the new technology.

In the computer-controlled environment at Faircrest, the human–machine interface was located primarily at control stations rather than at the equipment itself. The pulpits (control stations) were glass enclosed and situated away from the actual process operation. When these stations were being designed, there were no production operators on hand, so it was necessary for the manager–engineers to design this interface. These people understood the technology and were able to take the role of the operator at simulated controls. In several cases, cardboard mock-ups of a pulpit design were created to test the design. From the pulpits, operators would be able to monitor and control the process by way of numerical data, graphic displays, and closed-circuit TV. Traditional electromechanical control panels would be backed up by computer controls. Computer controls were designed to make the transition between the two control systems as easy as possible. Single-key commands were common. The primary function of the operator was to be a steelmaker, not a computer operator, and the controls were designed with that in mind.

The operator was to take an active part in running the process. The function of the computer was not to make all decisions automatically and reduce the operator to a process monitor. The function of the computer was to provide the operator with the information necessary to perform his or her job. The operator was to receive information from the process, from the material, and from algorithms that would suggest the optimal processing parameters. Based on his or her knowledge of steelmaking, the operator would have full authority to accept or reject the solution, to ask for alternative solutions, or to override the system completely. The human interface with the computer went beyond the ergonomic concern of reach, placement, and presentation of visual displays. The computer became a decision-support tool that was consistent with the company policy to allow the operator to have control over the process and to provide the resources with which to do it.

The process design path and the human resource path of activity recombined just prior to the time of personnel mobilization. By that time it was clear how the equipment would operate and how the operators would function. The first group of people to come on board were the "maintainers," as the mechanical and electrical maintenance people were called. These people actually helped with the final stages of equipment installation. Although there was not a significant opportunity for the maintainers to provide input to the equipment design, at least the design would be clearly understood. These maintainers did have an opportunity to test the pulpit mock-ups and, in some cases, went to an equipment supplier's site to check out equipment.

The first operating crew was on-site for training two months prior to start-up. Later, multiple crews were hired and trained. The start-up effort during this period was extensive, with 127, 000 people-hours of training. The topics covered: the Faircrest operating philosophy, teamwork, process technology, equipment-specific operation and preventive maintenance. The training effort was led by the Steel Project Team, which had been designing and building the processes from the inception of the project. All of those people who participated in the design and installation phases were expected to pass their knowledge on to those coming on board. From managers to operators, the writing of the training manuals was a collaborative effort.

Training tasks differed from the norm because the various operating groups had been selected and trained as crews rather than as individuals. All operators in a crew were supposed to be able to perform any task for which the crew was responsible, so all crew members had to be trained for all required tasks. Training took a "ground-zero" approach; that is, the first task was to overcome (unlearn) all bad habits and negative attitudes picked up during prior experiences in traditional job environments. Exposure to desired attitudes and knowledge followed.

As more and more was being learned about the processes and about the information and control system, training was increasingly being done by whoever knew the most about whatever was involved. Several operators contributed material for training manuals. It was also found that even though all of the maintenance people who had come into the Faircrest project had had apprentice training, they were not well enough prepared to satisfy the maintenance needs at the new plant. They needed extensive training to maintain this new equipment.

Performance

The first melt from the operation was poured on August 5, 1985. The result was salable product, which was quite rare for a start-up operation. Usually the first melt would be scrapped. By all project and production measures, the Faircrest plant was a success. The project itself came in $50 million under budget and two months ahead of schedule. The steel produced exceeded industry standards for dimensional tolerances, surface finish, and purity levels. The chemical composition of the many different alloys was one process target. Only 3 of the first 125 heats had to be scrapped because of the wrong chemistry. As of mid-1987 over 3000 heats had been poured without a miss. The design capacity was 12 heats per day, and in 1988 the plant had demonstrated the capability of 14 heats per day. Productivity was measured in labor-hours per ingot-ton and energy use per ingot-ton. The Faircrest plant operated at less than two hours of labor per ingot-ton compared to an industry average of almost seven hours per ton. Electricity usage was 22 percent less than at other Timken locations, and gas usage was 27 percent less. The "once-through" process that reduced the need for buffer inventory, and the process consistency that eliminated the need for rework, resulted in a throughput time of 24 to 48 hours, depending on the product. In the eyes of Timken management, this project had met and exceeded all expectations for success. When asked what was responsible for this level of success, the response was, "We did a lot of little things right."

HUMAN FACTORS CONSIDERATIONS DURING SYSTEM DESIGN

Overview

The Timken staff was well aware of our research interest in what human factors were considered in the design process. They expressed some concern that the success of the Faircrest project might be attributed solely to the human factors policies, while overlooking the many other design considerations involved. The technical achievement in this case was undeniable. The company set out to accumulate the best technology in the world and to integrate it with an outstanding computer system. They had achieved this goal. But along with this effort, concern for human factors did play an important role. In examining a list of human factors policies that might have an impact on the design process, we found that there were both explicit and implicit policies in place at Timken. There was abundant evidence that these policies were followed.

Operator Control

The policy on operator control was implicit—allow the operator to have general access to controls for process control. At the Faircrest plant the computer controls had manual overrides, so the operator could use his or her own judgment in operating the process. This was part of the design criteria and also part of the changed operating philosophy. There was a conscious and sustained effort to push decision making to the operator level. Computer systems were installed to support that decision making. In practice, however, some of the control systems worked so well that there was little need for operator intervention, but the opportunity to do so when necessary continued to exist.

Modification of the Operation

There was an implicit policy that directed the system design to allow modification within prescribed limits. The operator was to be given control to override the system as needed. If, based on his or her experience, the operator felt that the system could be improved by being modified on a permanent basis, this would probably have been done in conjunction with computer-support personnel. There was a constant challenge to keep the computer system up to date. The process was described as being a marriage of the computer and the technology. Systems were modified as required, so the operator did not have to circumvent them.

Feedback

There were explicit policies requiring that operators be given feedback on process variables as part of the decision-support function of the process control computer. There was also the requirement that equipment performance and product information be provided for operator use. In practice there was an abundance of process information feedback and visual feedback given to the operator. Anything that supported the operator's ability to do the job was given to him or her. In addition to process feedback, such business information as material and operating costs were given to the operator. Traditionally the operator would never have seen these figures.

Design Involvement

There was an implicit policy that encouraged the operator to make general suggestions during the design stage. In the case of the Faircrest plant, there were no production operators on-site during the design stage. Logistically that would have been difficult to arrange. Hiring production operators two or three years in advance of production start-up was not a viable option. What the company did do was select engineers who knew the process, technology, and operation to become process designers.

Safety, Health, Comfort, Stress

The company had explicit policies that exceeded OSHA standards on health and safety issues. There were implicit policies for operator comfort and stress. Maintaining equipment safety standards was the responsibility of those designing the plant and equipment. There were detailed specifications and extensive design reviews by both Timken and Dravo. OSHA requirements were covered at that time. The processing areas were clean, by steel mill standards. Most of the workstations were in glass-enclosed areas that were further guarded against accidental damage by nearby equipment. Safety programs were monitored on an ongoing basis. Workstations were redesigned when experience had shown that discomfort or stress existed.

Maintenance

The requirement for operators to perform first-level maintenance was covered by explicit policy. Equipment was designed or selected to facilitate this. Because of the once-through processing concept and elimination of redundant equipment, maintenance was a critical issue. With no provision for rework and no backup equipment, it was essential that everything be done right the first time and that no equipment breakdown be able to hold up production. This fairly simple concept had far-reaching implications for both equipment and organizational design.

According to the company's operating philosophy, the team concept in place at Faircrest required that "regardless of individual job descriptions, all members of a production team assist in returning the facility to operation in the event of a breakdown." To accomplish this, several elements had to be in place: first, the equipment design had to be easy to maintain and repair; second, information had to be available to diagnose problems; third, employees had to have adequate training to be able to take appropriate action; fourth, there had to be no organizational constraints, such as restrictive work rules or job classifications, that would prevent such action from occurring; and fifth, employees had to have the sense of ownership and motivation to want to return the operation to production.

Postinstallation Worker Participation

The Faircrest plant operated in a mode of "continuous improvement." One of the engineers remarked, "In the design phase there comes a point when you must freeze the design and go with it." The implication was that there would always be more to be done to improve the process. Timken had an implicit policy encouraging operators to suggest process improvements. Three years after the beginning of production there was ample evidence that operators were providing that input. In one case, a processing step was added to improve the quality of the steel. In another case, operator-recommended changes to a computer display permitted better control of the scrap distribution going into the furnace. In a third case, controls on a shearing operation were modified for more efficient handling. Not only were these improvements suggested by the operators, but Timken made resources available so that prompt action could be taken to implement them.

Ergonomics

There was an implicit policy that required specific design review for ergonomic concerns. In the design of the pulpits that were to be the primary workplaces for those controlling the processes, much consideration was given to the physical interface of the worker with the control and monitoring equipment. Cardboard mock-ups of workstations were built to test such items as reach, seating, line of sight of the process, and placement of monitoring devices. There was also ergonomic consideration of the information-processing aspects of the job. Displays were in graphic form whenever possible, so the operator could instantly assess process conditions. These were backed up with alphanumeric displays that provided additional detail. Switching from one mode to another required only single-key manipulation rather than extensive keyboard input. The responsibility to check the ergonomic design lay with the design team.

Training and Skills

In general the Timken skills policy was to design equipment to the existing skill level of its employees. At the Faircrest plant, however, the combination of new technology, the design process, and the new organizational philosophy led to the need for a significant amount of training and skills upgrading at all levels.

Although the steelmaking process essentially remained the same, the equipment and control systems for each of the process steps were all of advanced design. These high-tech systems had to be learned thoroughly by a few key people and then this knowledge transferred to all people working in that area. The computer system was designed specifically for the Faircrest plant, so this was another technology that had to be learned by the teams and then transferred to the operating employees. Not only did the technology have to be learned, but the skills of developing teaching plans and writing training manuals also had to be learned.

Members of the design team were chosen both to engineer this project and to remain with it as operating staff. This meant that the technical people would have to learn the managerial skills appropriate to operating this facility. New college graduates were hired to be part of the project team and eventually to become shop supervisors. These people had no supervisory experience, so all of these skills had to be taught to them. This meant that a significant amount of managerial and supervisory training was required. Technical knowledge was gained through their assignments to design teams for different areas of the plant.

Skilled hourly maintenance people were recruited from the ranks of Timken journeymen or from the company's apprenticeship programs. Even with this training behind them, it was necessary for these people to have extensive training in the new technologies. Many of these people came on board early in order to assist with vendor preoperational testing and installation of the equipment.

Aided by a no-strike agreement with the union, Timken management was able to apply stringent criteria in selecting the operating personnel. They were able to search outside the company to find a sufficient number of qualified candidates. Of the operating employees at start-up, only 20 percent were from other Timken plants, the remainder were new hires. These people were also brought in early to learn specific skills, the quality and metallurgical aspects of the job, company goals, the Faircrest operating philosophy, and an introduction to the computer systems. Ongoing training programs continued to enhance employees' abilities to function in the multiskilled environment.

Summary

From a human factors standpoint we found that there were many explicit and implicit human factors policies in place that had an impact on the design process. Although the training manager expressed a regret that human factors concerns and operator involvement had not been brought in earlier in the design stage, it appeared that this might have been very difficult to do from a staffing standpoint. In any event, there were few, if any, serious detrimental effects on the design or process technology as a result of the timing. It was very clear that there had been a top management desire to make radical changes in human factors policies from the very inception of the project. The design of that philosophy evolved in much the same way that the plant technology evolved. Timken had demonstrated that it could "do many little things right" to achieve its state-of-the-art technology and to match it with a state-of-the-art human factors environment.

IMPACTS OF THE PLANT DESIGN

Overview

The Faircrest project was a success in the eyes of management. It was, in fact, considered somewhat of a coup that a relatively small, traditional company could succeed in such a bold and ambitious endeavor. The people who participated in the project considered it to be an all-time career "high." It was described as an exhilarating experience, providing more organizational freedom than had ever been known. The operation continued to improve beyond the design specifications. With all of this, it was difficult to see a downside to the project, but management continued to address two concerns: (1) how to maintain the level of commitment and enthusiasm in the Faircrest organization, and (2) how to assimilate the technical staff back into the mainstream Timken organization.

During the design and construction period, activity had been at a fever pitch with long hours of overtime, weekend work, and travel. This caused some strain on both employees and their families. For those who were still part of the Faircrest plant at the time of this study there was some relief that the period of intense activity had ceased, but there was also an expression of concern that some of the vitality of the plant might be fading. There was definitely a hope that the good that had been achieved would not be lost over time. There was also concern about a "we–they" attitude that had developed between Faircrest and the rest of the company. While this apparently posed no serious problems, the Faircrest identity was there: "It was like being singled out for special treatment."

For those who transferred back to the main office, there was the shock of returning to a more hierarchical and bureaucratic world. Signatures, approvals, and cost justification once more became a way of life. This adjustment was particularly hard for some of the new college graduates whose first job had been the Faircrest project. There were some who could not adjust to that environment; they left for firms such as Dravo where they could repeat the Faircrest experience at other companies. Timken obviously did not want to lose these people in spirit or in fact. It continued to struggle to find ways to maintain and encourage the motivation and drive that was exhibited at Faircrest.

There were no reports of problems at the union leadership or operator levels.

Corporate Management

The obvious risk that the Timken corporate management had been willing to take to invest so heavily in the Faircrest Steel Plant, the high level of financial support provided to the project, and the trust they showed in giving the design team so much autonomy indicated management's commitment to the project and its willingness to try nontraditional creative approaches. As a consequence, a much keener interest in the FSP project emerged than would normally be expected, and this apparently resulted in a fairly close relationship with the team. On the basis of the success of the project, corporate management stated that the design process followed at Faircrest was to be the model used for similar future projects.

From a competitive standpoint, Timken's management had succeeded in establishing a world-class steelmaking facility whose product exceeded industry standards for dimensional tolerances, surface finish, and purity levels. Yet the average number of labor-hours per ingot-ton was less than one-third the industry average. The venture was reported to be highly profitable.

Design Team Managers

Under the design team concept all decisions were arrived at and implemented by mutual agreement of all team members. Each decision was made at the lowest feasible level, so the role of the design team manager became largely that of a facilitator. When the plant went into full production, each team manager changed to "area manager," but the decision-making philosophy of the project carried over into production. Each area manager had broad latitude to manage. A manager was expected to continually study and monitor the technology in the team's area of responsibility and to make certain that all team members understood the technology fully. Because design of the manufacturing processes was not static—both operations staff and operators were always seeking better ways to do things—the process control system had to change as well.

Design Engineers

Initially the major concern and challenge for the design engineers had been to design machines and other equipment that could operate continuously without breaking down. The team concept, however, generated ideas from many sources, including other design team members, operating staff, and supervisors. The design engineers had the responsibility to help develop these ideas and to determine their cost. Although they adapted readily to this role, a number of the design engineers had great difficulty in accepting the human resources aspects of the equipment and process designs.

Design Team

The pressures of the project had their impacts on the design team. Trips occurred with little or no advance notice. Long hours and weekend work disrupted family life. Each of the team members with whom we discussed this admitted how difficult it was on all concerned, but all said (some with hesitation) they would do it again if the opportunity arose. The sense of satisfaction gained from the tremendous responsibility each team member carried and the feeling of belonging to a group of this sort somehow made it all worthwhile.

Supervisors

First-line supervisions in both production and maintenance became directly involved in the process and system design at the drawing review stage. The production supervisors also handled most of the plant start-up. They were a major source of human factors input to control system and process designs, most of which involved computer interfaces. The supervisors described their role as much more technical than ever before, as indicated by their involvement in process and system designs. They were also often involved with the operation of the computer systems and frequently had to analyze a considerable amount of data. Because operators were responsible for correcting their own mistakes, supervisors were deeply involved in training and mentoring roles that required close communication with the operators. They often had to train the operators to be comfortable with the computers and automation, and let them know what automation could and could not do. Instead of doing things themselves as they had previously done, supervisors provided the information and tools that enabled operators to get the job done. Many supervisors, however, had a difficult time adjusting to this mode of operation.

Operators

Although most of the operators had not been selected or identified at the time the processes were designed and, therefore, had not contributed to the initial design, they soon became completely in charge of each heat. Operators felt a sense of ownership in the process and, in particular, in their pulpits with the computer aids and control devices. They were encouraged to use their own judgment and experience in their use of information provided by the support systems and to recommend changes in the information that was presented, the controls they used, and even the pulpit arrangements and layout. Because these people now had computer feedback and aids, they began to reduce what had been art to science. Also, because the operators were fully responsible for their own jobs, they were expected to get whatever resources were needed and do whatever was necessary to perform the work.

A gain-sharing plan for the operators was installed as a recognition and reward for the contributions of the operators.

SUMMARY

The Faircrest Steel Plant project and the events leading to its inception have had major impacts throughout Timken. Although corporate management and those involved with the planning, design, start-up, and operation of the Faircrest Steel Plant were affected most directly, the overwhelming success of what in 1982 had seemed like a very risky venture brought Timken favorable attention from the nation's financial and business communities. Laudatory articles in *Iron Age* and *Management Information Systems Week* portrayed the company as being willing to do whatever was needed to be a leader in its field in terms of stability, product quality, and price. Company management felt that potential customers had begun to look at Timken with respect and with assurance that the company would be around to serve them for a long time. Management also felt they had shown American industry that you must at times take steps that cannot be justified by return on investment alone. The FSP project had shown that good investments can pay off and that company managements can be motivated to back projects like it. Of paramount importance, however, was the fact that Timken had gained a seven-year competitive advantage on the entire low-cost alloy steel industry in terms of technology, process design, organization, control, and management expertise.

REFERENCE

MCMANUS, GEORGE J. (1987 August). "The Faircrest Plant Fulfills Timken's Hopes," *Iron Age*, pp. 22–28.

8

WESTINGHOUSE ELECTRICAL SYSTEMS DIVISION FLEXIBLE MANUFACTURING SYSTEM

THE COMPANY

The Westinghouse Electric Corporation was founded by George Westinghouse in 1886 as The Electric Company, a conglomeration of the 12 companies he had founded over the previous 10 or so years. One of the first and best known was the Westinghouse Air Brake Company, maker of safety air brakes for railroad locomotives. Several of the original 12, such as the Westinghouse Elevator Company, were still maintaining their autonomy 103 years later as units of the Westinghouse Electric Corporation.

Through the ensuing years Westinghouse Electric's business activities continued to diversify such that, at the time of the visit of the study team, they were major participants in the areas of broadcasting, defense electronics, electrical machinery and systems, financial services, beverage manufacture and distribution, residential and commercial real estate, elevators, furniture systems, watches, transportation systems, refrigeration systems, hazardous waste treatment and disposal systems, printed circuit boards, electronic products, and energy and utility systems.*

Westinghouse Electric's 1987 sales and operating income was $10.7 billion and net income was $0.7 billion. Approximately 45 percent of the company's sales and operating revenues and almost 50 percent of its operating profit were generated by its activities in energy and advanced technology. A major portion of these activities involved Defense Electronics, the segment of the corporation that included Westinghouse's Electrical Systems Division (ESD) in Lima, Ohio, the site of our case study.

*Dates of on-site visits: January 9, 10, and 13, 1989. The Electrical Systems Division was acquired in 1992 by Sundstrand Corporation, and became part of Electric Power Systems, a unit of Sundstrand Aerospace.

The study involved a new flexible manufacturing system for the production of aircraft generators. Westinghouse and General Electric almost completely dominated the market for these generators, with market shares close to 40 and 60 percent, respectively. In the broader field of defense electronics, Westinghouse had been running about seventh in sales volume. Market competition in the defense area was described as "fierce," largely because of the rapid technological advances that were being experienced in this area. The market had been growing, with the Department of Defense as the major customer.

A five-year summary of sales and operating revenues and net income (after taxes), and a three-year summary of cash flow from operations are presented in Table 8–1.

The Westinghouse corporate offices were centered in Pittsburgh, Pennsylvania. Their many plants, offices, and sales outlets, however, were scattered throughout the United States. Many were located offshore.

The Electrical Systems Division

Westinghouse first came to Lima, Ohio, in 1936 when it established its Small Motors Division there. An Aircraft Department was added during World War II to develop and produce electric devices for military applications. These included power-generating equipment, motors, and dynamometers for military aircraft. Because of growth in the business, the department developed into the Aerospace Electrical Division and eventually was renamed the Electrical Systems Division. It designed and manufactured electrical power generation, control, distribution, and conditioning equipment for space, airborne, surface, and undersea vehicles. Their electrical power systems were used on the majority of military aircraft in the free world, on most of the operating Boeing and McDonnell-Douglas civilian craft, and on the European Airbus. The Division also had facilities in Los Angeles and in Santa Isabel, Puerto Rico. The mix between military and civilian business, on a long-term basis, tended to a 50-50 split. The total number of employees in the Lima plant was 902. Annual sales were in the range of $100 to $150 million.

The ESD General Manager and his management team were all located at Lima. As shown in Figure 8–1, he reported to the Vice-President and General Manager, Marine and Electrical Systems Divisions, who was located in Baltimore. The ESD Operations organization is shown in the lower part of Figure 8–1.

Table 8–1 Westinghouse Sales and Operating Revenues, Net Income, Cash Flow from Operations. 1983–1987 ($ millions)

	1983	1984	1985	1986	1987
Sales and operating revenue	9,533	10,265	10,700	10,731	10,679
Net income	449	536	605	671	739
Operating cash flow			781	1,026	322

Source: Westinghouse Annual Report, 1987

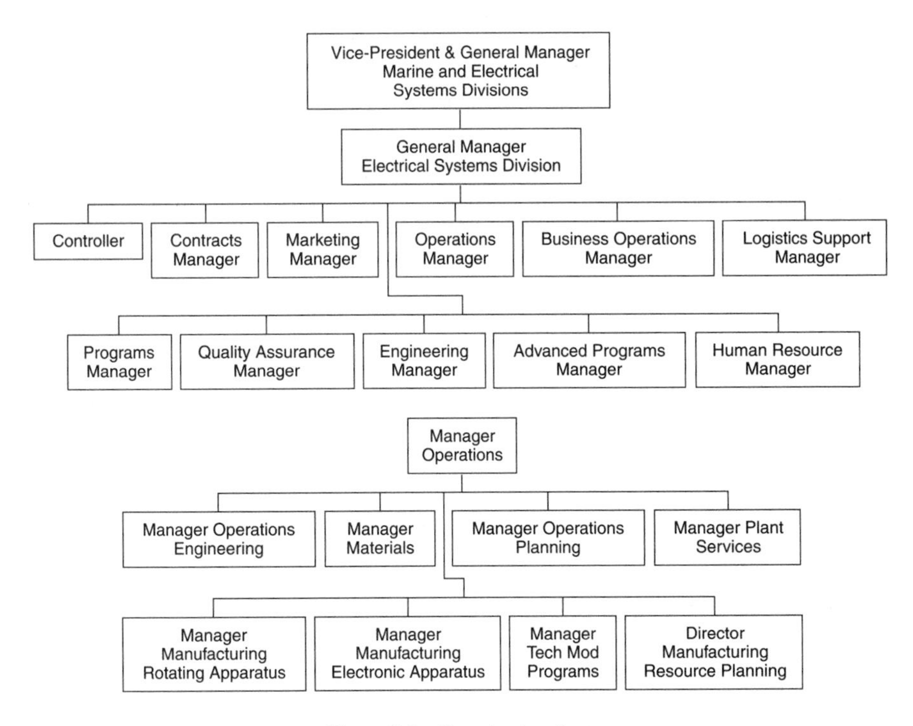

Figure 8–1 Organization charts

As part of the General Dynamics Industrial Technology Modernization Program, ESD had begun an extensive plant and equipment updating and modernization effort. Support for this came, in part, from the Department of Defense Industrial Modernization Incentive Program. At the heart of ESD's modernization under this program was the design and installation of a Flexible Manufacturing System (FMS) to replace most of their antiquated machine shop.

THE PRODUCT

In the late 1940s, the aircraft industry standardized on 400 hertz (Hz; one Hz = one cycle per second) AC aircraft power generators. Westinghouse continued its major role in this market, supplying electrical generators for military aircraft including the B-52 and commercial aircraft such as the Boeing 700 series. The latest technology for aircraft electrical generators in 1989 was the Variable Speed Constant Frequency (VSCF) system, which ESD produced for many of the most advanced military and commercial aircraft. Some electrical generators from ESD were used in tanks, submarines, small military naval surface craft, and space vehicles.

The smallest generator from ESD was rated to deliver 5 kilovolt-amperes (kva) and the largest was rated at 600 kva. ESD's most commonly produced generator was rated at 60 kva, large enough to power approximately 100 room air conditioners. Generators with the same power rating were custom designed to meet different specifications and to allow them to fit in the space available in different aircraft.

In both the military and commercial aircraft markets the key to product success was to capture the contract to supply generators for the first production of a given aircraft. Once an airplane was in production, aircraft manufacturers generally stayed with the same generator supplier for the life of that aircraft. Price competition was so strong that it normally took eight years to pay back the costs for a new generator. ESD was successful in this environment, with over 75 percent of commercial jet aircraft using their electrical power systems and components.

ESD supplied generators for aircraft being newly manufactured and spare parts for aircraft originally equipped with ESD generators. A specific aircraft design might continue in use for as long as 40 years, so the product life of an aircraft generator could be just as long. Approximately 25 percent of ESD sales was for spare parts.

All of these features of the aircraft electrical generator market combined to require ESD to produce a wide variety of generators. Often customers ordered only small quantities of generators at any given time.

Role of the Flexible Manufacturing System

ESD used its Flexible Manufacturing System to machine the parts that went into their generators. Because generators were rotating devices, most of their parts were cylindrical. Two-thirds of the parts were made from solid, round, steel bar stock. Most of the rest were cast parts that were roughly cylindrical. Most of the parts ranged in length from 1 to 14 inches and varied in diameter from 1 to 12 inches. Typical machining tolerances were ± 0.0005 inch, with special tolerance requirements up to ± 0.0001 inch.

At the time of the site visit, the FMS was being used on over 300 types of parts needed by ESD.The machines in the FMS had replaced most of the machine tools at ESD. A small number of parts were machined by ESD on the few old machine tools that were not removed from the plant when production shifted to the FMS. Parts that were not produced in the ESD Lima plant were subcontracted. ESD engineers pointed out that this approach differed from that of many factories that continued to use their old machine shops along with a special-purpose FMS designed for certain families of parts.

Relationship of Product Features to FMS Capabilities

Product features that were most relevant to the design of the FMS included: (1) long product life cycle, (2) need for only small batches of parts at any one time, and (3) wide variety of parts.

The long product life cycle and production of small batches meant that knowledge about making the part could easily be lost between production runs of a specific part. Programs used by the computer numerical control (CNC) machines in the FMS were a very efficient means of recording and retrieving that knowledge. The up-front cost of preparing the CNC program and supporting documentation for an operator could be spread over the lifetime of the product.

Computer control of the FMS machines made it faster and more economical to change them from producing one part to producing another. This flexibility was desirable for the wide variety of parts that ESD used.

Small-batch production of a wide variety of parts required more movement and storage of work-in-process parts than production of large amounts of uniform items. Five of the machine tools in the FMS could perform both turning (removing metal from the circumference of a rotating cylindrical part) and milling (putting holes or grooves in a part). This combination of machining capabilities within one machine tool reduced the handling and setup requirements for parts. The integration of the Automatic Storage and Retrieval System (AS/RS) in the FMS was intended to make the handling, storage, and control of generator parts more effective and reliable.

THE FMS FACILITY

The purpose of the FMS was to convert raw material into finished parts and record production information about the parts. The major items located in the FMS area included:

- An Automatic Storage and Retrieval System (AS/RS)
- Two horizontal machining centers with multiple-pallet storage systems
- Four turn/mill machining centers with carousel parts feeders and robotic parts loading/unloading devices
- One turn/mill machining center with bar feeder and carousel parts feeder and robotic parts loading/unloading device
- One small bar machine with bar feeder
- One coordinate measuring machine

A schematic layout of the FMS is shown in Figure 8–2. A description of the components and their mechanical support systems follows. Information about computer control and support of the FMS is presented later in the chapter.

The AS/RS was built in a straight line through the middle of the FMS. ESD used it to move material between machines and to store work-in-process. The AS/RS had 1860 storage locations. Two conveyors at each machine delivered material and tools to the machining stations and returned completed parts and tools to storage in the AS/RS. The AS/RS also had a pair of conveyors serving the tool preset station and a pair of conveyors for material receiving and finished parts exit. Some parts were occasionally moved on hand-pushed racks rather than by the AS/RS. All of the FMS machine tools were adjacent to the AS/RS, placed like ribs around an AS/RS backbone.

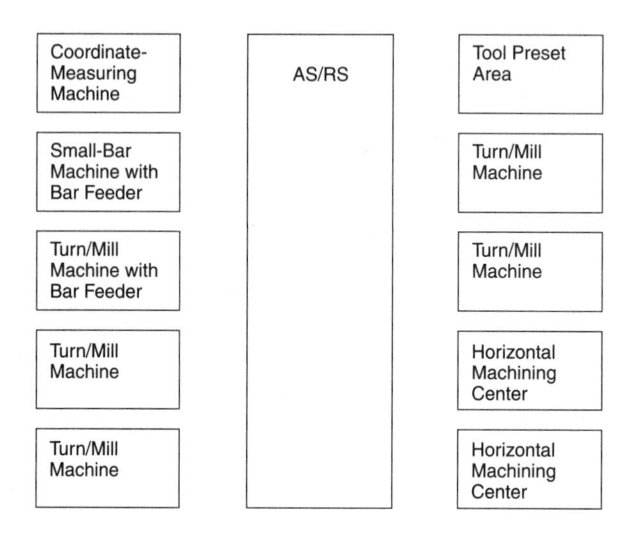

Figure 8–2 ESD Flexible Manufacturing System layout

The horizontal machining centers (HMCs) were CNC machine tools that worked on the larger parts such as generator frames and end bells. There were two of these machines located adjacent to each other. A Toshiba HMC was the first machine brought in to pilot-test the FMS. Toshiba declined to meet the same specifications for a second HMC, so the second HMC was supplied by Sajo. The HMCs used multipallet parts feeders and fed the parts to the machine using automatic pallet changing. A comparison of the capabilities of the two HMCs is shown in Table 8–2.

All five of the turn/mill machining centers (TMC) were identical except that one TMC had a bar feeder. These CNC machine tools could perform both turning and milling. The performance and features of the TMCs are shown in Table 8–3.

The small bar-machining center used a bar up to 12 feet long as its raw material to make small parts. The small bar-machining center made parts faster than the other

Table 8–2 Horizontal Machining Center Specifications

	SAJO	TOSHIBA
X-axis travel	31.5″	31.5″
Y-axis travel	27.5″	27.5″
Z-axis travel	25.5″	25.5″
Positioning accuracies over full travel on any axis	±0.0004″	±0.0004″
Continue horsepower	20 HP	20 HP
Tool capacity	123	90
Pallet capacity	7	11
Machine-mounted probe	Yes	Yes
Machine control	A.I.I. Producer	A.I.I. Producer

Table 8–3 Turn/Mill Machining Center Features

X-axis repeatability	± 0.0002″
Z-axis repeatability	± 0.0002″
Maximum turning diameter	21.25″
Maximum distance between centers	61.02″
Automatic tool changer with 60 tool magazine	
Automatic chuck jaw changer with 15 sets of jaws	
Totally programmable tailstock with a 5″ quill stroke	
Totally programmable steady rest with a 15–115mm range	
Total tool management package	
Automatic work piece measurement and compensation	
Automatic tool touch measurement and compensation	
Automatic work piece loading/unloading robot with a 40-kg capacity	
3000 rpm spindle with a 40-HP drive	
3000 rpm rotary tool with a 7.5-HP drive	
360,000 points addressable in C-axis	
Torque monitoring of both drive spindle motors	

machining centers, but it was able to work only on parts with diameters of less than 1.5 inches. Table 8–4 shows the features of this machine.

Two chip-recovery and coolant-recycling systems supported the FMS. Chips are waste metal removed from the parts by the machine tools. The chips were flushed from the machine by directed streams of coolant or moved by conveyors inside the machines. Chips exited from the machine into flumes in the floor where coolant was sprayed to flush them to a pit that was emptied by an inclined conveyor. The chips were carried on the conveyor to metal bins, which were removed from the FMS by forklift trucks. Both the chip-recovery system and the coolant-recycling system used flumes cut in the floor of the FMS. The flumes were covered with removable metal plates that were even with the floor surface.

Machines on one side of the AS/RS used oil-based coolant and were served by one coolant-recycling system. Machines on the other side of the AS/RS used a water-based coolant and were served by a separate recovery system. Parts made of magnesium alloys were machined on the machine tools that used the oil-based coolant. Many of the steel parts were made on machines that used the water-based coolant.

New tools to replace worn tools were kept in the tool preset area. Tools were placed into mounting fixtures that could be loaded into the machines. Completed tool

Table 8–4 Mazak Small Bar Machining Center

Features	
6000 rpm spindle	DNC capability
1/4″–1 1/2″ bar capacity	Tool touch probe
1000 ipm rapid traverse	Work piece measurement
Fanuc 10T control	

and fixture assemblies were measured to determine how far the assembly deviated from the specification for the assembly. These "offsets" were recorded on forms that accompanied the tool assembly to the AS/RS for storage and subsequent movement to a machine tool. The offset data were used to adjust a machine to the particular measurements of the tool when the tool and a machine were being set up to run a job.

During January 1989 the coordinate-measuring machine in the FMS was being prepared for use during part production. ESD personnel anticipated that it would help control part dimensions. The coordinate-measuring machine was a computer-controlled machine that measured the dimensions of parts accurate to ± 0.0001 inch. It automatically selected a part for measurement from either of two positions on a conveyor and prepared a written report of several hundred measurements in 15 to 60 minutes. This rapid, detailed, and precise information was expected to promote fast and accurate indications of undesirable trends in quality.

Comparison of FMS to Former Machine Shop

The former machine shop consisted of stand-alone machine tools that had an average age of over 30 years. The newest machine in the shop when the FMS came on line was 15 years old. Some of the machines were nearly 45 years old.

The FMS improved efficiency primarily by reducing the number of steps required to make a part. Table 8–5 shows examples of two production processes used to make the same generator frame. The first example (labeled "Before") shows the traditional process steps prior to the FMS. The second example (labeled "After") shows the processing steps used by the FMS to make the part. The number of separate operations was cut from 15 to 7, and the number of moves of the part from 17 to 4. Table 8–6 is a summary comparison of the number of steps required to make two other parts by traditional process steps and on the FMS. Although not the case in these two examples, in some instances the actual metal-removal time of the new machine tools was longer than the time of the old machine tools, but total process time by the new equipment was less.

Introduction of computer control to ESD machining operations brought with it a requirement that the machine tools be taken out of production to be used to develop the software to describe the machining operations for a part. The software, or part programs, were developed by manufacturing engineers with help from operators. Parts produced by a program had to be approved by quality engineers before the

Table 8–5 60-kva Generator Frame Process Steps

Part		Operations	Moves	Queues (delays)	Store	Inspect
Generator frame	Before	15	17	18	2	4
	After	7	4	7	2	2

Table 8–6 Other Parts Process Flows

Part		Operations (steps)	Moves (steps)	Queues (delays)	Store (minutes)	Inspect (minutes)
Shaft	Before	15	16	32	34.62	2.918
	After	5	6	5	1.47	1.225
Stub shaft	Before	25	25	24	13.63	2.259
	After	12	12	12	4.99	1.433

program could be used for routine production. At times, because of the many operations required to complete one part, program development for the part could be carried out for a few hours a week each week for up to several months. The total machine time to prepare a part program could vary from 1 hour to 100 hours. At the time of the research visit, the machine/operator time used for part program development was not severely limiting machine/operator time available for production. The difficulty was to find a convenient time when a machine, its operator, and a manufacturing engineer were all available to work on a part program.

Computer Support

The FMS Control System computers, FMS machine tools, and ESD business systems were designed to be interconnected through a communications network. This architecture allowed production orders and engineering information to be sent to the FMS Control System, which was expected to prepare appropriate information for the FMS. Table 8–7 shows the major functions of FMS computer support. All of the functions planned to be a part of the FMS had been tested in situations involving the FMS Control System's host computer and an individual machine tool. Not all of the planned functions of the FMS Control System were in place as of January 1989.

The Quality-Control function provided for collection of quality data from the machine tools by the host computer. At the time of the visit, this information was not processed after collection. The personal computers for statistical process control (SPC) that were described earlier in this report were not connected to the FMS Control System, and the reporting function had not been prepared for use. Preventive maintenance information was provided weekly.

Development effort during January 1989 focused on providing automated access to the Control System by all machines and the AS/RS for use of all planned functions as needed. Some of the machines had automated access to all planned functions, but true integration of all machines with all computer-based functions was incomplete. The FMS was being operated essentially as a collection of individually automated devices that were interconnected by activities of the ESD staff.

Table 8–7 Major Functions of FMS Control System

Order Control	Distributed Numerical Control
Entry	Store
Tracking	Download machine code (MCB)
Production Planning	Reporting
Scheduling	Production reports
Exception reports	Management reports
Schedule and Dispatch	Operating Interface
Miniload	Data entry
Computer numerical control	Operating aids
Monitor	Tool Control
Order start	Tool status
Process start	Tool offsets
Process Planning	Preventive Maintenance
Detail tasks	Maintenance data
Routing	Maintenance schedules
Quality Control	
Collect and distribute part data	

Role of Machine Operators in the FMS

A brief overview of planned responsibilities for various groups of employees who would be involved with daily operation of the FMS was written in an ESD document titled, "FMS Operation Philosophies." The document described the expected responsibilities of machine operators in the FMS as:

- Setup and operation of multiple machine tools
- Presetting of tool assemblies
- Miscellaneous deburring and miscellaneous operations
- Part quality
- Daily preventive maintenance as directed by the FMS supervisor
- Review, evaluate, and utilize Statistical Process Control (SPC) data

ESD management wanted operators eventually to carry out all of these responsibilities. At the time of the visit, (after 18 months of FMS operation) operators were not yet fully involved in all aspects of these tasks. A description of actual operator responsibilities and progress toward the above goals follows.

By January 1989 the FMS had been developed to the point that operators were setting up machines in response to written orders to perform operations on a part by:

1. Obtaining written documentation of the operation and part quality requirements. This assembly of documents was called a "packet." The packet was intended to instruct the operator on the details needed to make the part.
2. Getting the lead operator for their side of the AS/RS to enter a request into the AS/RS computer for material and tools.
3. Loading parts into the machine's parts feeder and tools into the machine's tool feeder.
4. Transcribing tool offset data from forms accompanying the tools into the machine controller.
5. Starting the program.

After setup, each machine ran while being observed by an operator. This was different from the original plan, which called for the operator to run several machine tools at the same time. Several labor issues pertinent to this situation were under review by labor and management at the time.

Presetting of tool assemblies was done by an operator who was not scheduled to operate a machine tool at the same time. This scheduling practice also represented a change in the original plan wherein each machine operator was to have set all of his own tools. Quality-control inspectors verified tool assemblies before they were placed in the AS/RS for use in production. Deburring was also performed at a workstation dedicated to that activity.

Operator activities in quality control involved hand gauges and fixtures to make measurements for entry into the SPC personal computers in the FMS. Once the coordinate-measuring machine had been programmed to measure enough parts, it was expected that the operators would use it and spend less time using hand gauges and fixtures. Operators adjusted the machines to eliminate undesirable trends in part quality.

Preventive maintenance involved lubrication and cleaning. ESD management wanted machine filter changes to be done by operators, but filter placement in the machines required operator activities that were restricted by labor agreements. At the time of the research visit, operators were performing lubrication and cleaning of accessible machine parts, but maintenance workers were removing machine covers to change filters. Management and labor representatives were reviewing labor agreements related to this issue.

Part Quality

Given the critical role played by generators in the proper functioning of aircraft, part quality was an essential factor in process design. The most obvious design features of the FMS that were expected to contribute to quality of parts included: (1) computer numerical control machine tool repeatability, (2) Statistical Process Control methods used in the FMS, and (3) part- and tool-probing features of the machine tools. In this regard, the FMS was a success. Engineers who designed the FMS reported that the expense of scrap material produced by machining error at ESD declined by 70 percent during the first 18 months of FMS operation.

Computer control of the FMS's machine tools and the quality of the tools made the machines' actions precisely repeatable so that dimensions of the parts from the machines were consistent. When a machine tool had been set up to make a part, all subsequent parts in that batch had almost precisely the same dimensions.

Statistical Process Control was implemented in the FMS by personal computers installed for operator use. Operators were to enter measurements and review trend graphs of the measurements in relation to acceptable limits. Operators were expected to adjust their machines to correct undesirable trends and to inform their supervisor if a SPC computer gave them information that bad parts had been made. Measurements entered into the SPC personal computers by an operator went into a database that could be used for analysis and planning by manufacturing and quality-control engineers. An ESD quality-control engineer reported that this database helped him do quality analysis studies more rapidly.

The quality-control data that operators were required to collect were described to operators in documents included in a packet with the part program. Operators were expected to assure that parts met all quality aspects described in the quality documentation accompanying the part program as well as other quality aspects described in more general-purpose documentation.

The machine tools included probes that gave information about the position of the tool and part in the machine. The probes were used to assure operators that they had properly set up or adjusted a machine tool. The probes also informed the part program of the orientation of the part for automatic part positioning before the program began commands to cut the part. These uses of the probes reduced error. The probes had the potential to make quality measurements automatically, but at the time of the visit this aspect was not totally developed, so process quality-control measurements were done manually.

Relationship of the FMS to Other Stages of Production

The FMS carried out the machining stage of the production process for ESD products. Material used in the FMS was inspected prior to loading it into the AS/RS to assure that it met quality requirements. After a part went through all required machining operations, it left the FMS for storage, assembly into a completed electrical device, or shipment to customers who ordered spare parts.

The FMS was the first of several production operations in ESD to be automated. Automated part heat treating was in operation during the research visit. Automation projects that had been initiated, but not completed at the time of the research visit, included improvement of assembly operations and installation of an automated guided vehicle system for material and product movement. During January 1989, early stages of installation for the improved assembly operation and early stages of installation for the automated guided vehicle were visible. ESD management was also in the process of developing new labor agreements and relationships with its operators to raise the performance of the division.

THE DESIGN PROCESS

Background

The decision to automate the Lima plant was a strategic one. ESD's manufacturing facility had been held in very low regard by its customers. In an age in which high-technology parts were being required by both military and commercial customers, the operation and atmosphere of a 1950s-vintage factory was not very reassuring. It sent out a message that cast many doubts on the technical expertise of ESD and its ability to keep up with both present and future demands. In order to preserve business, ESD had to bring its factory up to date and that, for all practical purposes, meant automation of the machining operations. ESD also wanted to let both its employees and its customers know that it was going to stay in this business for the long haul.

For the approach to the machining operations, the selection of the turn/mill centers and horizontal machining centers represented the most advanced replacement for the existing numerical control (NC) machines. By combining these machines with distributed numerical control (DNC) from a central host computer, ESD would also be able to obtain a tighter control of the machining operations. Skilled machinists were becoming very hard to find. Quality and consistency considerations were cited as the prime reasons for wanting to establish tight control over the process. The plan for DNC processing of the parts programs would return this control to the engineers.

Better control of materials and tools was the motivation behind the installation of the AS/RS. This, along with a central tooling preset station at which the tool setups were calibrated as they were being kitted for specific jobs, was intended to limit the substitutions or adjustments that the operator could make on the job. Just-in-time (JIT) operations were planned for the future at ESD, so ready accessibility of the tools would be essential to make that operational. Flexibility of operation had also been a key design criterion. If changeovers could be simplified, production lot sizes could be reduced and less inventory carried.

The motivation for the total system design was to build for the future. Equipment was specified to include advanced features and options. Systems were designed so that advanced features could be added later. For example, a broadband computer network was installed so additional peripheral equipment could be added as the need arose. Even if these features were not to be used immediately, the potential for growth was designed into the system.

The plan of ESD to build a "new factory in an old shell" was implemented in a three-phase program. Phase I, factory analysis, was done in conjunction with General Dynamics' Technical Modernization program and funded as part of the government's Industrial Modernization Incentive Program. This program provided incentives for contractors to modernize their operations with the objective of long-range price reduction on products purchased by government agencies. The study looked at the "as is" conditions of the factory. ESD personnel were trained to conduct in-depth analyses and assessments of factory operations. From this study 11 potential projects were identified as candidates for factory improvement. Of these, two proposals were approved for seed-money funding by the Air Force, with General Dynamics' Industrial Technology Modernization Program serving as the administrating agent. Nine other projects were funded with Westinghouse money under the same program.

Phase II covered the detail design and planning phase of the project. It included prototype development and site preparation. Design of the FMS took place during this phase of the program.

Funding was allocated to the projects in Phase II under a Memorandum of Understanding with the government. ESD management reported that, "Such an instrument must be negotiated with mutual trust of the parties. Although there are targeted savings associated with the modernization projects, there is also a recognized risk with regard to their success or failure. The company and the government would be sharing the risk as well as the benefits. "For this particular FMS project the agreement allowed Westinghouse to keep an adequate return on investment for 10 years. The government would benefit from the reduction in product cost.

Phase III of the program was implementation, or release to production—a fulfillment of the commitments made under the Memorandum of Understanding. The release date was very important because at that point the sharing of cost savings with the Air Force was to begin. Once the FMS was implemented, both parties were eager to share the benefits of the factory modernization.

The FMS installation at ESD was just one of the projects under the Industrial Technology Modernization Program, but it was one of major impact. In addition to its high cost, it was a significant advance in the technical expertise of the Lima plant. The objective of this installation had been not just to bring the plant's 1950s technology up to date, but to leap ahead to a 1990s technology. Spanning this technology gap was a major undertaking for those involved in the project.

Project Participants

The design of the FMS technical team was guided by a Planning Task Force led by a Program Manager. The task force was made up of 12 individuals who represented the following general functions: product design, quality assurance, manufacturing engineering, management information systems, contracts, production operations, maintenance, and accounting. These were organizational groups that would have to provide input to the design of the FMS or to integrate aspects of it with their own functional responsibilities. The technical team was also part of the task force.

The FMS technical team was organized in 1984 after funds were actually made available to the project. Late in 1984 three people were hired to be responsible for FMS equipment, software, and industrial engineering. Eventually seven people were assigned to the team. Some contract help was also added. Software support came from Westinghouse operations in Pittsburgh. At one point, at least one of the programmers did work at the ESD site, but for the most part there had been a long-distance relationship between the technical team and the people developing the software.

The product design group had a relatively active involvement with the FMS design team, particularly with respect to new part design, resulting in an increased effort to improve part design for manufacturability. This new emphasis required a significant increase in coordination between the product design group and the manufacturing group.

The quality assurance organization was represented on the task force for the FMS. Its role, however, was not frequently mentioned as being an integral part of the technical team. Quality issues were certainly not overlooked in the design of the system. There were plans to install a coordinate measuring machine, to include Statistical Process Control in the computer system, and to make measurement of the parts integral to the machining cycle. The quality organization also had a very active role in qualifying the new processes once the machines were on-line. What was reported as having been missed in the original design process was a close integration of the daily quality assurance tasks into the work design, work flow, and layout of the FMS. Later efforts to bridge this shortcoming involved use of an Automatic Guided Vehicle system to move parts between the FMS and inspection.

There was an attempt to involve some of the production work force in the project. This consisted primarily of communication and training rather than active participation by operators or supervisors in equipment or system design. Quarterly meetings were held with the machinists for communication purposes. Six operators, a shop supervisor, and the union president were among those mentioned as being involved with the introduction of the FMS. The union president was also among those who visited the equipment supplier plants in Kentucky and in Japan. One operator did the first-piece runoffs at the Kentucky site.

Three formal training programs were conducted to prepare the operators and supervisors to operate the FMS. The first was given by an instructor from a local two-year technical college and covered general machine shop practice and protocol. Those who successfully completed this portion of the training then received a machine-tool-specific, eight-week course taught by a professor from a university located 60 miles away. The first six operators who successfully completed this second course received six weeks of comprehensive on-the-job training from the responsible ESD manufacturing engineer. These six operators then provided the on-the-job training for the rest of the operators.

The technical team reported having great difficulty in soliciting input for the FMS design from other functional areas in the plant. They felt that most people lacked both an understanding of the impact the FMS would have on their jobs and the time to thoroughly learn what the system had to offer.

Design and Installation

There were three distinct lines of FMS design activity associated with Phase II of the program: design of the bar-machining cell, the casting-machining cell, and system integration with the AS/RS. Although these had been defined as Stages I, II, and III, they were done concurrently rather than sequentially to assure compatibility of the entire system. This overlap of activities meant that there were many different pressures competing for the attention of the team members. Apparently there were demands for support of the day-to-day production activities in the plant as well. This diluted and fragmented the attention of the FMS team, and team members responded by spending many hours beyond the normal workweek at the plant. In discussing this project, management acknowledged the team members' dedication and effort. The rewards came later. All three of the original team members were promoted into management positions, a move, however, that served to diminish the continuity of technical expertise within the team.

A similar sequence of design activities recurred for both the bar-machining and casting-machining stages. The preliminary work to prepare these operations for transfer to the FMS involved establishing part families for all 3000 parts supplied to the new and secondary markets. These part families were then analyzed to determine what operations would be done in machining cells. The part configurations, in many cases, were complicated. Parts that were turned and generally cylindrical might also have tapped and drilled holes at various angles or have machined surfaces or keyways. The turn/mills had to accommodate these operations in a variety of sequences. This flexibility had to be designed into the machines. Each configuration would have to be programmed in detail, so there was much to be gained by establishing part families requiring similar operations. If related parts required the same operations, dimensional differences could be handled by the computer-managed machines. The establishment of these part families was a time-consuming task, but one that was considered necessary and worthwhile.

Once the turning and machining activities for representative families had been established, the next step was to determine what sort of equipment was available to do the job. The design team visited vendors, attended many trade shows, and gained access to other companies that were willing to discuss their experiences with FMS installations. The union president was included on some of these trips. After these investigations were complete, the team was able to specify the type of system required for ESD. Machines were specified that would combine both turning and milling operations, thus eliminating material handling and setup steps associated with transferring parts back and forth between single-function machines. ESD also wished to have one uniform machine tool control system. Suppliers found it difficult to respond to these requirements. Domestic companies were unable to respond at all; the supplier of the prototype machine essentially said, "No, thanks, one controller conversion was enough." ESD was forced to restudy the availability of suppliers and eventually to accept two different control systems on the machines in the FMS.

To integrate the machines into the total system design, an FMS Operating Philosophy was formulated in March 1985. This provided the conceptual framework for operation of the entire FMS system. The technical team reported that this was developed primarily as a guiding document for the software developers. It also included requirements for quality control, business systems, operator responsibilities, materials handling, maintenance, tool management, design engineering, and job descriptions for all who were to work with the system.

Part programming was a detail-intensive activity that started before delivery of the machines. After the machines were built, there were other tasks to be completed, such as first-piece runoff tests at the vendor site, delivery, installation, training, debugging, and release of the system to production. The release-to-production date for any given machine was the point at which that machine attained a 25 percent load on a two-shift operation. Part programming would be continued well beyond the release-to-production date and was to become a part of the ongoing activity of bringing outsourced parts back into Westinghouse.

The integration of the system was based on the operating philosophy developed around the design of the machines. Included in the system integration was the AS/RS, a major piece of equipment that had to be tied to machine operation. It stored the tools and raw materials that supplied the machines. It also moved parts to the coordinate-measuring machine and subsequent operations. The software for this and for much of the system was written at Westinghouse's Automation Division. A close association had not been developed between the technical team and the Pittsburgh software people, and this made it difficult to communicate on common terms. The software people were often not aware of the coming requirements of the numerically controlled machines. This made later trouble-shooting efforts very difficult. If they had it to do over again, many on the design team indicated that they would develop much closer working relationships with the software people at an early point in the design process.

Problems Encountered

One of the design objectives of the FMS was to use only proven technology. The designers did not want to endure the sort of problems associated with the introduction of equipment or systems that were still in the test or development stages. The designers tried to identify and anticipate problems, but problems still arose. Some issues were resolved, but others had not come to a satisfactory conclusion by the release-to-production date.

The main problem with the operation of the FMS was that it operated as a group of eight distinct machines rather than as an integrated system. At the total system release-to-production date in December 1988, many of the machines were still operating in a manual data input mode rather than by having one part of the system talk with another. In the manual mode many of these stations, especially the AS/RS, were not very "user friendly." Operators had to load computer programs into the turn/mill machines in much the same way that they had loaded paper tapes on NC machines. The coding language used by the software designers was unfamiliar to the ESD people, and the software people did not really understand machine tools, so communication between these groups was difficult. There was a strong emphasis on quality at the plant, but the quality systems were not fully integrated into the FMS. The tools necessary to do the job were there, but the logistics of moving parts to be measured were awkward. The measurement probes within the machines were being used to measure tooling positions for setup, but the piece-part measurement function was still under development.

The FMS design team was very much aware of these technical problems and of the problems caused by their approach to system design. They reported many examples of what might have been done differently and compiled a list of the lessons they had learned. Their discussions with regard to these lessons learned gave no indication, however, of the magnitude of the problems or what had been the impact on the implementation of the system. Some of the problems were of the nuisance variety, but others appeared to be fairly significant. The lessons learned did highlight the team's many insights that came from the experience of designing a complex automated system. The team stated that these insights were applied successfully to subsequent installations at the ESD site.

Lessons Learned

The team members listed a number of lessons learned—actions and strategies for future projects that would be different from what they had done on this project:

- The relationship with the vendor is a key consideration. Specifications must be quantifiable. Terminology must be understood by both vendor and customer.
- Available software and controls were not capable of supporting FMS as ESD wanted it.
- We underestimated the cost of software; it is just as expensive to change software as it is to change hardware. Software people are good at software, not manufacturing.
- We would not have done the simulation. It was too painful to create a realistic model. The database was inadequate to feed the simulation program. Simulation would have been more useful once the system was implemented; then it could have suggested and evaluated potential design changes.
- Have a long-term contract with the vendor so that all of the machines are the same. The vendor changed machines over time so earlier machines are slightly different in operation from later ones.
- Have a dedicated design team with the same people throughout the project. If you change team members you lose continuity.
- Involve the operators on the shop floor; include the entire group that will interact with the system. Run this as a design team.
- Try to reassign a few seasoned machinists to work full-time as part of the design team.

- Ideally the designers should live with the project. When a similar project was done at another location, the project leader became the manufacturing manager.
- Get supervisors involved.
- Involve specialists, get to the functional people. It may be incredibly difficult to get input from these people because of time pressures and firefighting. We had to draw suggestions out of people. Many people can't visualize a concept on paper.
- Try not to be a guinea pig for new equipment; use only proven technology.
- Get out of the planning stage as quickly as possible and get on with the implementation.
- People need the positive reinforcement of seeing the project take shape.
- Don't always replace people with automation. People have pretty good "computers" in their heads.
- Don't overextend; don't use the worst case or most complicated process as a pilot.
- Management must also see success.
- Use a phased implementation. Debug the software on one machine; then train operators; then install the rest of the machines.
- Make software design specifications easy to read and understand.
- Establish a close working relationship with the software people early in the project.
- Document things that are learned about the system, particularly for troubleshooting.
- Get maintenance involved in the design stage, they have something to contribute.
- Build a sense of urgency toward the machine; have support people dedicated to it. It isn't just another machine.
- Bring the quality organization into the picture early. The coordinate-measuring machine should have been the first piece of equipment on the floor.
- Keep people growing with new technology as it develops. Don't wait for long periods before adding new equipment. Avoid the culture and technology shock.
- Supervisors must change the way they deal with people. They must involve people in decision making, get inputs, participation. At this point, not all of our supervisors are comfortable with this change.

ESD Evaluation of Success

In spite of some continuing issues with the FMS as a whole, there were significant gains realized at the time of release to production. The run time on the machines was 85 percent compared to 40 percent previously. The process cycle time was reduced by 50 percent. The number of operations was reduced by 50 percent. Machine uptime went from 50 to 90 percent. There was a 60 percent reduction in scrap rate on first-piece setup and an eventual 80 percent reduction was expected. Running scrap was reduced 85 percent. Overall quality was improved. There were better outcomes on complicated parts. Some of the complex electronics parts could be made in-house for less cost than when outsourced. ESD had previously not even had the capability of producing some of these parts. On the basis of these measures, the FMS was considered a success.

The production measurements at the time of release to production showed that the FMS was successfully producing parts at improved levels of productivity. The perceptions of the overall success of the project, however, appeared to be mixed. The people at ESD were obviously pleased at having come so far so fast with their concept of building a new factory in an old shell. The FMS was an ambitious undertaking and there was justifiable pride in what had been accomplished. There was no question that the objective of the FMS, that of getting parts through the system more productively, was being realized. There were frustrations, however, because the system was not complete and not all of the design features were operating as planned. This sense of frustration appeared to be greater at the lower levels of the organizational structure. At the top-management level there was an acknowledgment that the system was not yet functioning as designed, but this concern was not strongly voiced. Much more emphasis was placed on the positive aspects of what had been accomplished in spanning the technology gap. Middle managers, especially those who had been promoted from the design team, were closer to the problems. They were more candid about the difficulties and better able to stand back and assess the causes of the difficulties. The FMS had been a major learning experience for this group. They were in a position to take the lessons learned and apply them to new projects, so for this group and for upper management the FMS was a positive experience.

Most of the evidence of lingering frustration came from those who were still closely associated with the day-to-day shop-floor activities. Much of it seemed to relate to the lack of systems integration. The people felt that they had been pressured to get the machines into production just as soon as they had been delivered. This push to get each machine individually into production had detracted from the team's ability to make the total system operational. A general perception of those close to the shop floor seemed to be that top management did not fully understand the magnitude of the problems and the amount of work yet to be done on the system just to bring the FMS up to the design objectives. Regardless of these perceptions, ESD was organizationally committed to the project, as shown by the existence of a special engineering-support unit dedicated to FMS completion.

HUMAN FACTORS CONSIDERATIONS DURING SYSTEM DESIGN

The FMS was designed to facilitate material movement, and the machines were selected and installed to be as automated as possible with little need for operator intervention. The changed work role of the operator was envisioned early in the design stage, as evidenced by the operating philosophy established for the project. As listed earlier, operator responsibility was to include part quality, setup and operation of multiple machine tools, presetting of tool assemblies, deburring and miscellaneous operations, routine preventive maintenance, and statistical process control. Responsibility for designing the automation capabilities and operator interfaces lay with the software designers.

The design team reported having little interaction with the operators during Phase II, the detail design and installation period of the project. The team seemed to be in agreement that the operators would not have had much to contribute to the effort; the technology was too advanced. Quarterly meetings had been held with the operators, but in retrospect the team agreed that this was not enough. The shop's response to the question of participation in the design of the system was that it could have made constructive contributions if it had been introduced to the system from its inception and had been allowed to live and grow with it.

At the same time that the FMS was being designed, a plantwide restructuring of jobs at ESD was underway. One hundred job titles were reclassified into 28. Seven career paths were created. The path that included the jobs of the machinists who would be operating the FMS went from 30 job descriptions to one career path with three intermediate steps. In addition to this, the plant went from an incentive pay system to measured daywork. Advancement in the new pay structure was to be based on the acquisition of new skills through training. These changes were to have been in place before installation of the FMS. The negotiations with the unions, however, had been protracted, so the machines were already arriving on-site before the issues were resolved. The consequence of this was a perception of coercion felt by the operators. While the job reclassification did not seem to have an impact on the FMS design, it did have an impact on the operators' willingness to go along with the team concept and with tasks, such as the performance of first-level maintenance, that had been specified in the operating philosophy.

The technical team reported mixed success with the effort to consider some of the human factors in the design of the system. Policies and procedures were in place to assure attention to safety factors. While operator participation in the design of the equipment was largely absent, there was evidence that operators did have an opportunity to influence changes to the system after it was installed. Although control of each machine was narrowly limited to minor tooling adjustment, workers were encouraged to broaden the scope of their activities through a team concept of work, preventive maintenance of machines, and statistical process control. Extensive training on the operation of the new machines was provided. The new compensation policy made skills upgrading part of the machine operators' career path, so continued training was assured.

Operator Control

Westinghouse had an explicit policy that restricted operator control of the process to clearly defined parameters.At ESD the operator control of the FMS was limited to the very narrow scope of changing tool offsets. The magnitude of the tooling offsets was determined by a tool-setter using precise measurement devices at the tooling preset station. Then, when the tools were loaded into the tool carousels, these established offsets would be manually entered into the machine controller by the operator. If, during routine operation of the machine, piece-part measurements indicated that the offsets should be changed, operators did have the capability to change them, and these changes would be uploaded onto the computer tape.

Modification of the Operation

Although the project had an implicit policy to design systems so that operators might modify them within prescribed limits, there did not appear to be much opportunity for this to occur on the FMS. The operators were not permitted to write or change part programs other than to adjust the tool offsets.

Feedback

The machine controls and the computer system of the FMS did provide information that the operator could use to adjust the process. The computer screen at the machine had several options that the operator could access to get information about the operation currently being done on the machine. Some of the displays were in machine code. The two operators interviewed by the research team indicated that they knew how to read and interpret at least some of these codes. Computer-based statistical process control charts were displayed from data that the operator entered into the system. From this information the operator would be able to determine if the process remained in control. Charts of production output and quality data were displayed in a central area of the FMS. Other process manuals were also available for use in that area.

Design Involvement

The company's implicit policy regarding worker participation was to encourage the operator to make suggestions or comments during design activities. However, at ESD, involvement of the operators in the design of the FMS was negligible. The extent of involvement or participation with the operators appeared to be the quarterly review meetings. The engineers did not believe that the operators, in particular, would have anything constructive to contribute to the effort. They found that most operators could not envision alternative ways of doing their jobs or anticipate what benefits might come from the FMS installation. They felt that operators were also unwilling to spend the time to learn what the system could do for them.

The perception of those on the shop floor was somewhat different. These people believed that they would have been able to contribute to the design of the equipment if they had been afforded the opportunity to learn more about the system during its development.

Safety, Health, Comfort, Stress

Safety features were a prime consideration. The safety officer reported that hazards had not increased with the FMS installation. Engineers had extensive checklists to follow while designing equipment and layout requirements. Drawings of the systems were reviewed for safety and were signed off by the safety officer. In-plant safety inspectors and state inspectors also approved the installations. In addition to standard safety features, there were also those that were specific to the FMS application. The fire-extinguishing system that covered the chip and coolant system was a good example of this. The oil-based chip and coolant system contained highly flammable materials. The system was protected by a carbon dioxide delivery system that automatically activated to extinguish fires. The effectiveness of the system was documented by videotape. The general environment of the FMS workplace was clean, well lighted, and comfortable. By direct observation, the work itself did not appear to be stressful.

Maintenance

There was an implicit policy that equipment design should facilitate first-level maintenance on machines by operators. This was included in the operating philosophy of the FMS. The company clearly wanted first-level maintenance to be included in the expanding work role of the operator, but the operators had not yet accepted this responsibility. Some of this reluctance was attributed to the union environment in which each concession was used to negotiate a specific benefit. This environment also inhibited the implementation of a teamwork concept for the FMS.

Postinstallation Worker Participation

The implicit policy was that operator suggestions or evaluations should be encouraged after installation of the equipment. The company encouraged this type of input, especially through group meetings with supervisors. Not all of the supervisors, however, were comfortable in the new role of problem solving, coordinating, and facilitating. They were more comfortable with a "boss" role. It was difficult to get these supervisors to hold meetings with the work force. Individual engineers, on the other hand, reported good working relationships with operators in the identification of problems and suggestions for improving the system.

Ergonomics

The company's explicit policy was that there be specific responsibility for ergonomic design. Although ergonomic design was not mentioned as a particular design criterion for the FMS, there did not appear to be any violations of ergonomic integrity to the physical layout of the entire FMS. As most of the material handling was automated by conveyor or robots, there were no apparent problems with lifting or moving large parts. The FMS had undergone a series of design reviews, including a "red team" review by industry experts to verify the soundness of the design and to guard against oversights.

Skills and Training

There was extensive operator training associated with the installation of the FMS. ESD wanted to provide education in basic machine programming as well as skills training required to do the job. There were three segments to the program. The first was conducted in-house with a CNC desktop lathe. The operators used a personal computer to write programs for the lathe. The second was a six-week course at the Lima Technical College. This class provided instruction specific to the Mazak turn/mills that the operators would be using. The third was a six-week period of on-the-job-training.

Training on the FMS and throughout the plant was expected to continue. The General Manager of the plant was committed to having a well-trained work force This need to upgrade skills and education was a topic of discussion in all of our interviews. A shop-floor training center had high visibility. Human resource policies, such as the new compensation and recruitment policies, further reinforced the emphasis on technical education. The skill-based compensation program used skills training as the basis of promotion and pay increase. New recruitment policies favored those who had at least two-year technical college degrees.

IMPACTS OF THE SYSTEM DESIGN

At the time of the research team's visit, the FMS project had already had far-reaching impacts both within and outside the Westinghouse Electrical Systems Division. The fact that the ESD had taken the initiative to become involved with the Industrial Technology Modernization Program and had obtained seed money from the Air Force to fund two of its proposed projects provided the division with timely favorable publicity. Relations immediately improved with General Dynamics, who could anticipate future savings on equipment and parts for its aircraft through more efficient operations at ESD. This activity and the resulting potential for increased business convinced other customers that, in spite of the age and condition of the Lima plant and its equipment, Westinghouse was not going to close it.

Plant Management and FMS Management

The FMS project gained ESD considerable credibility with Westinghouse corporate management, who were persuaded to put more resources into ESD and to encourage the development of the Marine and Electrical Systems Divisions. This interest and support, coupled with the intensity and excitement of the project, were personally very satisfying for the plant management.

Because of the importance of the FMS project to ESD, the FMS management team felt pressured. They tried hard to meet the original schedule even though many factors affecting the rate of progress were beyond their control. They put in an extensive amount of unpaid overtime, not because the company required it but because the management team was excited about the project and wanted to be there.

Almost all decisions were made and implemented by mutual consent among the members of the management team, often in conjunction with other groups such as manufacturing engineering, shop-floor supervision, or quality assurance. Consequently, they had to develop much greater technical ability, people skills, and salesmanship. Some management team members experienced frustration in trying to get people to think in "tomorrow's terms," that is, to be able to envision where the project was heading so that "ownership" by all parties could be established.

Design Engineers

At Westinghouse, the title "Design Engineer" referred specifically to a product engineer. Since essentially all parts, including those for the newer Variable Speed Constant Frequency systems, were to be made on the FMS, design engineers had to design these parts to facilitate their manufacture on the FMS. They had to broaden their thinking regarding their designs. They had to interface much more with manufacturing engineers and adapt their part designs to the manufacturing system. ESD had already instituted a system of formal intermediate and final design reviews in which both product design and manufacturing engineering groups were required to participate and to sign off on all designs.

Manufacturing Engineers

As the complexity of the manufacturing technology grew with the development of the FMS, the manufacturing engineers had been called upon to fill more central roles in both system and product designs. They had to develop greater computer skills and expand their heretofore strictly mechanical horizons to become more systems, computer, and information oriented. As a result, they were having to look beyond their own islands of responsibility and consider the impact of design changes on the whole system. The manufacturing engineers' role in the FMS also required a much more frequent presence on the shop floor, so they relocated their desks and workstations there. Their areas of responsibility were also greatly expanded. One engineer, for example, was responsible for all five of the Mazak turn/mill machining centers and for writing part programs for them. Some manufacturing engineers still felt, however, that they should be handling more than facilitating and programming roles such as determining locations for work-in-process. Nevertheless, at least one of the manufacturing engineers described his role as both challenging and exciting.

A trend seemed to be developing toward more manufacturing engineering positions being filled by engineers with baccalaureate degrees rather than by people working their way up the technical path from the shop floor, perhaps with some technology training. In general, management seemed to be developing a much greater appreciation for the role of the manufacturing engineer than appeared to be true in the past. A new job classification called Fellow Engineer had been created as an additional step above Senior Engineer, and at least one such position was expected to be filled in the near future.

Training

The FMS brought with it the need for broad job functions and accompanying broad job descriptions for operators and supervisors. Immediate extensive training in many areas was a necessity. For plantwide training, Westinghouse allocated $6 million and got an additional $1.7 million from the State of Ohio. ESD also built a training center on the factory floor. FMS training was conducted by machine tool vendors and local university personnel, as well as in-house personnel. Those who did not want to enter training were allowed to sign a revokable waiver saying they were not interested and were content to remain in their current situation. It was felt at the time of the visit that the company had not yet obtained the full benefit of these training efforts. Reading skills on the shop floor also needed improvement because of the increased amount and complexity of written instructions, blueprints, and communications with the shop-floor computers. Selection and implementation of programs to address this situation, however, involved union negotiations that had not yet been completed.

Operators

The impact of the introduction of the FMS on the operators appeared to differ with the age of the person affected. Some of the older operators whose reading skills were weak faced the choice of having to improve their reading ability, attempt the technical retraining in spite of their reading disadvantage, or leave. Some in this group did, in fact, choose to retire. Younger operators who had two-year college or technical school degrees and the few who had four-year degrees were doing well. It was too early at the time of the visit to determine what the effect of the FMS would be on the younger operators who had no particular education beyond high school.

Many management and engineering personnel felt that the operators had received pressure from the union not to be overly cooperative with the FMS project activity, so the union could bargain for additional benefits as a consideration for getting the FMS project completed. Although the Manager of Human Resources Operations did meet with operators and group leaders for ideas, the design of the system was left to the technical people and suppliers.

Accident rates, which had been low before the FMS, appeared to be unchanged.

Union

The union obviously wanted the FMS project to succeed. However, there was a strong feeling among some members of the Planning Task Force and the Technical Team that the project had provided the union with a golden opportunity to "trade" each of the management's desired changes in rules or procedures for some increase in the union's benefits or bargaining position. Many of the operators, however, were interested in doing multiple machine operations and were apparently getting disenchanted with the union because of the delays that kept them from doing so. The union also raised the issue of protecting jobs by requiring all parts to be produced in-house regardless of the relative costs of doing this. The union finally agreed that some parts should be made outside if it were more economical to do it that way. In spite of some changes like this, union relations were generally perceived by management as a major roadblock to progress.

Supervisors

The effects of the FMS on the supervisors' jobs probably were more severe than on any other group in the plant. Their traditional role had been that of "straw boss," in which the supervisors told the operators essentially everything they were to do. By the time of the visit, their role had become basically that of facilitator, trainer, and mentor. The supervisors had to equip and encourage the operators to participate at both the technical and operational levels of the process. They had to keep operators informed and help them develop feelings of ownership of their jobs and the overall process. The major demand on supervisors had evolved into serving as sources of technical, process and production knowledge. During the first two years of operation of the FMS, the supervisors had gone through a difficult transition period under considerable pressure, during which they had to work longer hours to get their jobs done.

A different set of skills had become essential. An educational background that better prepared supervisors for the changing environment was receiving increasing attention in placement considerations. The last supervisor to resign, for instance, was being replaced by a person with a bachelor's degree in manufacturing technology.

Maintenance

In spite of being allowed unlimited overtime, the maintenance force was not able to keep up with the increased workload. Their work now included major factory modernization and upgrading projects as well as routine maintenance. They had to learn to take care of the new machines. This added load was creating a critical situation because the FMS did not always have a redundant machine available that could take over when a machine was down for service or repair.

In keeping with the FMS operational philosophy, ESD wanted the machine operators to perform more maintenance activities. They also wanted to have maintenance people as an integral part of the work teams. A satisfactory balance had not yet been worked out, however.

The "Old Work Force"

A major impact of the FMS was evident relative to older members of the work force, regardless of their particular positions. This group included operators, supervisors, technicians, manufacturing engineers, and quality assurance engineers. Most of them were the individuals in these and perhaps other job categories who came to work at ESD, perhaps were even sought out by ESD, a decade or two ago when experience and the willingness to work hard were the main requisites for hiring and advancement. They were used to working in a traditional machine shop, probably at less than peak performance, but in what had become psychologically very secure surroundings. The FMS forced them out of their "comfort zones." Concurrently, new employees, most with at least some formal education and some with college degrees, were being hired.

As reported to us by some of the supervisory and supporting staff, some of the older operators who had been in ESD most of their working lives felt that management was telling them they were no longer important. This feeling may have been reinforced by management decisions to have workers use computer systems in place of some of the personal skills required in the past.

In two or three of our conversations with operators, however, little was said regarding this subject. The operators in this group indicated that they were not resistant to the changes in equipment and technological updating, but were hurt by their apparent neglect by management. It was felt by some that this could have been avoided by involving this group in the up-front planning activity for the project. Then, when the equipment came on-line, at least some of them would have been comfortable with it, felt ownership for it, and been able to contribute to its functioning.

SUMMARY

From a human factors standpoint the FMS was designed to limit the operator's control of the process to a narrow range of functions. To replace the relative autonomy that the machinists had at their older machines, management had envisioned operating the FMS under a team concept that would provide the operators with a wider variety of meaningful tasks. Neither the design process used with the FMS nor the change to the new classification and compensation policy did much to support the shop's acceptance of this team concept. From the considerable expense in training, it was clear that ESD was committed to bringing its human resources into the 1990s along with the technology. It was also apparent that a change in organizational culture was going to require every bit as much effort as the change in technology.

9

WHIRLPOOL CORPORATION CLYDE DIVISION ASSEMBLY LINES

THE COMPANY

The Whirlpool Corporation was founded on November 1, 1911, as the Upton Machine Company by brothers Louis and Fred Upton with help from their uncle Emory. The brothers' first washing machines were designed by Louis and Emory Upton. The company began production in St. Joseph, Michigan.

A business relationship with Sears, Roebuck, and Company began in 1916 when Sears decided to market the Upton-manufactured washer. By mid-1925, the Upton Machine Company had become the only source of supply for Sears washing machines.

In a 1929 merger with another washer manufacturing company, the Upton Machine Company became the Nineteen Hundred Washer Company and acquired the brand name Whirlpool. The firm changed its name to Whirlpool in 1950. Since the name change, the corporation has produced washers and a variety of other appliances. Many of the new products were added after mergers; for example, air conditioners and kitchen ranges were acquired from a merger with two divisions of the Radio Corporation of America in 1955. At the time of our case study, the list of items produced by Whirlpool included washers and dryers for home use, coin-operated washers and dryers for commercial laundries, refrigerators, room air conditioners, food freezers, microwave ovens, ranges and exhaust hoods, residential ice makers, residential trash compactors, food waste disposers, dishwashers, dehumidifiers, vacuum cleaners, food mixers, food processors, and small hot-water dispensers.*

*Dates of on-site visits: September 14, 19, and 20, 1989.

The corporation administered a network of franchised service companies. The Whirlpool Acceptance Corporation was Whirlpool's wholly owned finance subsidiary.

Whirlpool's annual revenue in 1988 was $4.42 billion, and its net income was $161 million. Approximately 92 percent of Whirlpool's revenue and 90 percent of its operating profit came from home appliances. A five-year summary of net sales revenue, total revenue, and after-tax net income is presented in Table 9–1.

Three of Whirlpool's strategic business units and their 1988 sales were:

Kenmore Appliance Group, $1.44 billion.

Whirlpool Appliance Group, $1.678 billion.

Inglis Limited, $0.351 billion.

In 1989, the major domestic appliance makers were Whirlpool, General Electric, White Consolidated Industries (owned by AB Electrolux), Maytag, and Raytheon. GE was the largest competitor, with 1988 annual revenue of $5 billion from major appliances. GE had invested over $200 million annually in improved production technology during the late 1980s. Whirlpool's net property additions were $217 million, $223 million, and $166 million in 1986, 1987, and 1988. Both GE and Whirlpool had made major investments in joint ventures with European appliance producers. Beginning in 1989, the Sears "Brand Central" retail appliance outlets offered both GE and Whirlpool brand name products at the same location, which intensified the competition.

Whirlpool corporate offices were in Benton Harbor, Michigan. They had plants in Michigan, Ohio, South Carolina, Kentucky, Arkansas, Indiana, Tennessee, and Mississippi. In January 1989 Whirlpool entered into a joint venture with N.V. Philips, which provided the firm with a presence in all of western Europe. The joint venture had plants in France, Italy, Spain, Sweden, and West Germany, and accounted for approximately 25 percent of worldwide major appliance sales.

Whirlpool owned 49 percent of Vitromatic in Mexico and 72 percent of Inglis Limited in Canada. Other parts of the corporation had operations in China, India, Australia, and Brazil. There were plants in India and Brazil.

Table 9–1 Whirlpool Net Sales Revenue, Total Revenue, and After-tax Earnings, 1984-1988 ($ millions)

	1984	1985	1986	1987	1988
Net sales revenue	3,137	3,474	3,937	4,114	4,314
Total revenue	3,201	3,542	4,014	4,208	4,421
Earnings	190	182	202	187	161

Source: Whirlpool Annual Report, 1988.

The Clyde Division

The Clyde Division, in Clyde, Ohio, manufactured all of Whirlpool's domestically produced automatic washing machines. It was housed in a plant purchased from the Clyde Porcelain Steel Company in 1952 and in an adjacent plant purchased from Bendix in 1954.

The Clyde Division was a part of Whirlpool's North American operations and was in the network of manufacturing divisions that supplied products to the brand-oriented appliance business groups. The Clyde Division supplied washers designed by corporate personnel to the Whirlpool, Kenmore, and Inglis strategic business unit groups.

The top manager of the Clyde Division was the Division Vice-President. Reporting directly to him were managers for engineering operations, quality improvement, support operations, assembly operations, human resources, and finance. The staff for all of these positions was in Clyde.

Although almost all of the Whirlpool plants were unionized, the Clyde Division plant and plants in Marion and Findlay, Ohio, had no union.

Clyde Division management reported that three strategic decisions made by the corporation shaped their operations. These decisions were (1) to consolidate all automatic washer production in Clyde, (2) to produce a fundamentally new washer design featuring direct drive rather than belt drive, and (3) to upgrade production processes in the division.

THE PRODUCT

The new direct-drive automatic clothes washer, which was marketed under the brand names of Whirlpool, Kenmore, Roper, Inglis, and Kitchen Aid, was known within Whirlpool as the Laundry Engineering Advanced Product or LEAP. The product had versions with 24-inch or 27-inch cabinets for residential or commercial use. There were three basic capacities for wash loads. Other features were dictated by the mix of control console, dispenser, agitator, and basket. There were eight paint colors and two types of finishes. The Clyde Division classified their washers into 290 different stockkeeping units for inventory control and production-planning purposes.

The first version of the LEAP design was proposed in 1973. In 1979, a decision was made to build a pilot LEAP production line at the Clyde Division. Five hundred units from the pilot line were used for a customer field test in Florida in May 1980. Production was increased to 500 units per day in November 1980. The LEAP design was gradually introduced to more types of washers, beginning with machines with low sales demand and few features to allow a gradual increase in manufacturing process complexity. The latest variation of the direct-drive washer configuration was the Large Capacity Thin Twin. This combination product had a large capacity dryer placed above a large capacity washer in the same cabinet. The washer was shipped from the Clyde Division without a cabinet or controls to another Whirlpool plant for final assembly.

The LEAP design differed from previous washer designs because the basket, agitator, and pump were connected to the motor through a gear rather than a belt. The washer could be built from the inside out. Both features reduced manufacturing costs, increased water extraction from wet clothes in the basket, reduced the number of parts, and reduced energy needed for manufacturing. The improvement in water extraction occurred because the LEAP basket spun at 640 rpm compared to 505 rpm for the belt-drive model. The inside-out assembly sequence allowed for easier service in the field because the cabinet could be removed from the front of the machine and the machine could operate without a cabinet. Several features were included in the LEAP design to simplify installation for the customer. One of these features was that water supply hoses were attached at the factory. Another feature that simplified installation was that plastic restraints, which were attached in the factory to protect the motor from excessive movement during shipping, could be removed without removing the washer cabinet.

The belt-drive automatic washer and the LEAP were both designed for a useful life of at least 10 years. Some belt-drive washers remained in use for 30 years. At the time of the visit, less than nine years after product introduction, the actual useful life of the LEAP was not yet determined.

The LEAP's subassemblies included a gear/motor/pump unit, base and tub support, tub, basket, control console, top, lid, and cabinet. These subassemblies were made at the Clyde Division. The gear/motor/pump unit began with an electric motor. Clyde Division placed the pump and gear case parts onto the motor as it moved through a subassembly area. The base and the tub supports were made from stamped metal parts that were welded together. The stamping and welding was done at the Clyde Division.

Polypropylene for the tub was delivered as spherical pellets to a storage silo beside the plant. The pellets were pneumatically conveyed through rigid metal tubing to a hopper that automatically dispensed the pellets into 1500-ton injection molding machines. Some molding machines were unloaded by robots and some were unloaded by automated pick-and-place machines. Clyde Division engineers indicated that the pick-and-place machines, considered a type of "hard automation", were more reliable because of their simplicity. The only operators involved in tub subassembly manually loaded and unloaded equipment that heat-sealed a molded plastic dome (used to sense water level) to the outside of the tub.

The basket was made from three pieces of steel and tubing that were automatically welded together. The baskets were inspected and then the coating was applied.To further improve the balance of the basket, a water-filled plastic ballast ring was attached outside the top of the basket.

The top and lid were formed in presses and painted at the Clyde Division. Prior to 1981, approximately 90 percent of the tops and lids had a porcelain finish to prevent corrosion from water, detergents, and fabric softeners. Beginning in 1981, the Clyde Division began to substitute an organic powder coating for porcelain. By 1989, over 80 percent of the tops and lids were finished with the powder paint. The powder-coated tops and lids were finished with one coat that was 2.5 to 3 mils thick. The powder coating closely resembled porcelain in appearance, and was salt-spray and impact resistant. Powder coating required fewer steps to apply, had an application process that used less energy, could be applied by equipment that was less labor-intensive to maintain, and had safe emissions. The major disadvantage was that the galvanized steel used for the powder-coated tops and lids was more expensive than the steel used for porcelain-coated parts. The cabinet was loaded, formed, welded, and unloaded by a process that was unattended except for a setup person. There were three of these automated-cabinet weld lines that were simultaneously monitored by the setup person.

There were many quality inspections on the subassemblies prior to their use in final assembly. Statistical Process Control was heavily used by machine operators in areas that made metal parts, such as the tub support and agitator shaft. The dimensional quality of the base was checked by a laser vision system. The pump's leak test was automated. Baskets were automatically checked for concentricity needed for proper balance. The ballast ring was automatically tested for leaks before attachment to the basket. The finish of the tub and external subassemblies had extensive quality control.

Because the LEAP could operate without a cabinet, this made access to the unit easier when it was assembled and allowed more thorough inspection of a fully functional unit. The cabinet was the last part applied to the LEAP, so it passed through fewer points on the assembly line where it could be damaged.

As the process became more automated, it was found that the process equipment could not tolerate as much deviation in part dimensions as before. Part tolerances often had to be tightened to reduce chances that a marginally acceptable part would reduce the effectiveness of a high-speed operation.

All three high-speed assembly lines included several quality test stations. These included an automated water-fill test that tested the accuracy of the filling process, the quality of electrical insulation, and the integrity of the electrical ground. After the water-fill test, an assembler initiated a functional test, including checking water temperature controls, checking the pump-out process, and checking the control of the basket spin. After adding the top, lid, and cabinet, the unit entered a sound booth that automatically tested the unit for unusual noise. After the sound booth test, the unit went either to rework or continued along the line to be packaged.

Quality of the finished product was tested in several ways. A 2.5 percent sample of production output was sent to the division's Customer Acceptance Lab (CAL), where each unit was operated for 30 minutes. CAL testing included electrical tests, visual tests, and confirmation of correct packing of literature. Defective units were held for review by assembly line workers (called assemblers at the Clyde Division) who could correct the defect. After the assemblers' review, the defective unit was sent to a rework area for corrective action. A second off-line set of tests was done by the Reliability Lab, which tested two machines per line per shift. The Reliability Lab's tests included 72 hours of operation. A warehouse audit administered tests to washers that had already been put in cartons and moved to the division's distribution center.

To assure quality as changes occurred, the division had a management practice that dictated minimum levels of support for processes that had been designated as "critical." There were 42 "critical" processes in the summer of 1989, and each one required the following:

- A procedure to approve first-part quality
- A system for Statistical Process Control of the process
- Quality standards and verification methods
- Documented machine capability studies
- Up-to-date preventive maintenance
- Spare parts on hand.

To find product and/or process changes that would benefit the division, division personnel used Taguchi Methods to plan experiments. Changes were also solicited from employee involvement teams.

The Sears Roebuck Company recognized the quality of the washers produced at the Clyde Division. The division's largest customer had presented the division a quality award during each of the eight years that Sears had operated their quality monitoring process. Only six of Sears' 10, 000 suppliers had won eight consecutive awards. Other evidence of product quality was provided by comparing the rate of in-field service for washers made by different manufacturers, trends in reports from the division's Customer Acceptance Lab, and trends in warranty service expense. During the late 1980s, annual warranty service expense for the LEAP had fallen by 15 percent per year, and the rate of defects found in the Customer Acceptance Lab had fallen 30 percent from year to year.

THE PROCESS

To support the LEAP, Whirlpool invested over $155 million in plant, equipment, tooling, and engineering between 1979 and 1987. A large portion of the capital investment and all the tooling investment was for production changes at the Clyde Division. The new product required a new assembly process.

To maintain the production volume needed during the change from belt drive to LEAP, Whirlpool renovated only one of the Clyde Division's three high-volume production lines at a time. The time between installation and/or renovation of the three different assembly lines allowed improvement of the designs, installation plans, and operating practices. The first line that was redone for LEAP had the most automation and could assemble the widest variety of LEAP models. The later lines used automation only where it had been effective on the first line. The later lines were capable of producing only the regular and large capacity 24- and 27-inch models. One of the later two lines had been used solely for assembly of the 27-inch models.

The production process required a minimum of 140 assemblers for each assembly line. To maintain high-speed operation for such a large number of workers, control of the pace of the line had to be dictated by the process. Attempts to give assemblers time to correct defects as they occurred on the line resulted in excessive line stoppage, and this interrupted the rhythm of other assemblers and caused delays for all units on the line. Clyde Division personnel preferred that defects in the finished product be corrected in an off-line rework area.

The sequence for assembling major parts of the direct drive washer on the high speed assembly lines was:

- Start with base.
- Attach tub support.
- Add gear case/motor/pump unit.
- Add tub.
- Add and level basket.
- Add tub ring and agitator.
- Attach controls and connect them to sensors and actuators.
- Do water-fill test, check functions, and water temperature.
- Inspect during pump-out and spin.
- Add top, lid, and cabinet.
- Check for unusual noise.
- Add cardboard carton and stockkeeping labels.
- Move packaged unit to distribution center.

The washer was built from the bottom up and from the inside out while it moved on a pallet on the assembly line. The conveyors moving the pallets along the assembly line were a mixture of nonsynchronous and index-and-dwell types. The pallet moved to the next station when the index-and-dwell conveyor indexed or when the pallet on the nonsynchronous conveyor was released by the assembler. Manufacturing personnel preferred a uniform pace along the assembly line, which caused them to minimize use of nonsynchronous conveyors.

Operator stations on the high-speed assembly lines had a cycle time below 14 seconds. All stations involved very simple operations that were paced by the assembly line. For example, the sole function of one assembler at his station was to install one screw. Several subassemblies requiring longer than 14 seconds were added to the washer at two or three stations. Depending on the complexity of the model being assembled, each high-speed assembly line had a crew of between 140 and 205 assemblers. Each of the high-speed lines could produce 2,000 washers per shift.

Each of the three high-speed assembly lines had its own rework area. The first line that was redone for LEAP had a rework area with the capacity to fix over 150 defective units per shift. Design expectations that automation would maintain a low volume of rework were not met during early experience with the first high-speed

LEAP assembly line. Later high speed lines had rework areas with capacity to fix over 250 defective units per shift.

The Clyde Division assembled the mechanical elements of the washer portion for the Large Capacity Thin Twin on an assembly line that was separate from the three high-speed assembly lines. This fourth line was originally staffed with 22 assemblers, but when actual sales fell below projected volumes, the line was rebalanced for 19 assemblers. The line was called the High-Commitment (High-C) Line because extensive efforts were taken to encourage employee involvement. Operators on the High C-line performed more assembly at each station than was performed at stations on the three high-speed lines. Any assembler on the High-C line could stop the line to correct a defect. If the defect could not be fixed during a short pause on their line, assemblers from the High-C line reworked defective units in open areas beside the line.

Circulating overhead conveyors moved most large parts to the high-speed assembly lines. Parts that were delivered directly to the assembly line in this fashion included the tub support, tub, basket, splash ring, back panel, control panel, and cabinet. Tops and lids were moved on sectioned, wheeled carts that held one part per section to prevent damage to the finish. Forklift trucks moved boxes of other parts and materials.

The degree of automation of the four lines varied widely. Priority for automation was given to operations that required precise gauging, specific torquing, and heavy lifting, or that were hazardous. Automated gauging, for example, was used at the water-fill quality test station on the high-speed assembly lines and filling of the gear case with oil in the gear case/motor/pump subassembly area. The only assembler–computer interaction in these examples was during setup. Most of the automation for hazardous and/or heavy jobs was in parts fabrication or subassembly areas.

Automation was also used to implement a direct-flow supply concept and a computerbased material control system. The direct-flow supply concept synchronized part fabrication, part movement, and use of the part on the assembly line. For example, if the assembly line planned to change the color of the model being assembled, parts would leave the paint shop in the same sequence that they were needed and would move directly to the assembly line. Clyde Division engineers indicated that the combination of direct-flow material handling and good material control information could provide more flexibility in assembly operations, because changes from model to model could be done rapidly and efficiently. The Clyde Division used approximately 300 suppliers and had a goal to develop supplier relationships so that the number of suppliers could be cut to approximately 100. The largest volume of incoming material was steel and packaging. The division received daily shipments of these materials and turned over their inventory of steel and packaging in less than five days. The approximate daily value of internally made parts was $1 million. The division also made after-market spare parts for both the LEAP and belt-drive washers.

Control of production volumes of the various models was based on a schedule that set goals for production managers. The three high-speed assembly lines were operated two or three shifts per day for 250 days annually. Changing the number of shifts of operation of a line was usually done only once a year. The maximum combined capacity of the division's three high-speed lines was 18,000 units a day. Production of 14,000 units per day had been planned and carried out for several months for each of the years during the late 1980s. The High-C line operated for only one shift, with peak capacity of 250 units per shift.

All the high-speed lines had machine status displays mounted above the lines to provide process feedback. The status messages were less than 20 characters and were readable at distances of over 100 yards. The assemblers also received feedback on product quality in involvement teams, during defect reviews in the Customer Acceptance Lab, and in the division's daily newsletter. Feedback to assemblers on the High-C line came directly from Whirlpool's Marion Division, where the assembly of the Large Capacity Thin Twin was completed. Marion Division's feedback included future production volume needs and information on the quality of the assembly work done by the High-C assemblers.

Computer systems were organized in a four-level hierarchy. The first, or lowest, level used programmable controllers that interfaced directly to production equipment. The next level was called area control and was implemented with minicomputers or personal computers with multitask capability. The third level was implemented on the division host computer, an IBM 3090. This level had communication equipment typical of a mainframe environment. Processors on all three levels shared the division's four-channel broadband network. The division host communicated with the top level of the hierarchy, which was the corporation's mainframe.

The broadband network was installed throughout the division. It had communication channels for the Manufacturing Automation Protocol (MAP), a local area network, an Ethernet network, and a data highway. The data highway channel connected programmable controllers that counted production output on the assembly lines and the basket fabrication line with the schedule control system operating on a MicroVAX 3400 computer. The local area network channel on the broadband network was used by a computer-based system recording maintenance activities and by a computer-aided design system.The MAP channel of the broadband network was used by the time-and-attendance system and an inventory system.

As of summer 1989, the major computer-based applications that directly involved production personnel were the porcelain basket coating system and the time-and-attendance system. The porcelain basket system used automatic sensors and manual input to gather production and defect counts. The porcelain area control equipment included two DEC PDP 11/53 computers. Both of these computers were connected to the Ethernet channel of the broadband network, which allowed communication with the division host, the scheduling system, and a computer-based design system. The time-and-attendance system accepted input from time clocks and supervisors' terminals. Payroll processing was done on the corporate host, which communicated with the time-and-attendance system via the division host. At the time of the visit, assembly line workers did not use the computer-based systems at their workstations.

The division planned one addition to its manufacturing information system to support the direct-flow material control concept. Other computer-based applications would be installed only when the payoff for specific process improvements was seen to be greater than the cost of adding computer support.

The division employed approximately 200 hourly and 25 salaried personnel for maintenance. Maintenance Department records showed that there was no specific area or maintenance activity that consistently consumed large portions of maintenance resources. Maintenance personnel were dispatched from a central location. The dispatcher and maintenance personnel could communicate by two-way portable radios. The dispatcher used a computer-based system to keep maintenance activity records. There was no centrally located performance-monitoring equipment for production machines. Due to the short cycle times on the assembly lines, assemblers performed no corrective maintenance and did not do periodic preventive maintenance other than activities that could be included in their daily setup.

The division used written maintenance standards to guide consideration of maintenance issues during evaluation of new equipment. It was a division management practice to be sure that processes designated as "critical" had a preventive maintenance plan and that spare parts were on hand before the process was released for production use. More than 95 percent of the preventive maintenance activities for the "critical" processes had to be completed within a week of the scheduled start date.

THE DESIGN PROCESS

Planning for the manufacture of the LEAP line began in 1981. The product implementation strategy was to introduce the less complicated, low-end models first. Then, as production experience was gained, the division would introduce the more difficult-to-produce models with advanced features. Production of the older belt-driven models would be phased out as new model production volume increased. As the project developed, its scope was greatly expanded by the inclusion of more product models than had originally been planned. To meet this production increase, four assembly lines were required. Each line was introduced as a separate engineering project, and each project provided a learning experience for the next.

The four design projects for the LEAP assembly lines are described separately. Table 9–2 summarizes each line's main features and project timetable.

Whirlpool Assembly Line 3

Project participants. Manufacturing engineering responsibilities at Clyde were assigned to individual project engineers. These individuals were responsible for gathering the resources required to do the projects. Internal resources were available, particularly in the skilled trades, to help implement the projects. For the assembly lines, much of the equipment was built to specification by an outside vendor.

A senior project engineer was assigned to the first assembly line of the LEAP project in 1981. This highly experienced engineer was given total responsibility for the design of the manufacturing assembly process. On this line there had been con-

Table 9–2 LEAP Assembly Line Features and Product Timetables

Line #	Basic Technology and Level of Automation	Project Duration	Postimplementation Enhancement
3	Nonsynchronous, power and free line; high level of automation; assembler controlled line; extended spacing between workstations	Late 1981 Mid 1983	Mid 1983 Mid 1984
4	Power train drive; medium automation level; work zone stop control with automatically paced line restart; 3 unit buffers between work zones; workstations in close proximity	Mid 1984 Late 1985	Late 1985 Mid 1986
2	Drag line retrofit; index and dwell; low level of automation; workstation design allowed formation of social groups.	Early 1986 Early 1987	Early 1987 Late 1988
1	Very low line speed under assembler control; moved product only when quality was right; multitask workstations.	Early 1989 Mid 1989	Mid 1989 Late 1989

siderable collaboration with the product design engineers from the corporate design group. There was a willingness on the part of the product design engineers to modify the product where necessary in order to improve manufacturability.

Top management gave the project engineer complete support. The engineer was sure of management's absolute confidence in his ability to manage the project. It was a hands-off relationship except for the strategic design concept and periodic reviews. The engineer was given the charge to design for the "factory of the future," to provide complete flexibility on the line for all possible laundry models, and to use a Japanese approach to assembly line work.

Design and installation. The assembly line was fabricated and built by an outside vendor. The engineer chose the vendor who could provide the best resources with a minimum of bureaucratic red tape. The vendor would have to dedicate many of its engineers to the project in order to meet the project completion date, and it was essential that this commitment could be met. The Whirlpool engineer also wanted to be able to make on-the-spot decisions with the vendor's engineers without having to go through many layers of approval. The Whirlpool engineer described his relationship with the vendor as excellent. Segments of the assembly line were built and tested at the vendor site. Installation and debugging at the vendor site occurred during the first half of 1983.

The approach to the LEAP assembly line was to make it as automated as possible. Originally the line was to have had 30 automated steps, with the object of keeping labor content to a minimum. Line speed was to be 8 percent greater than older model lines. At the time of the case study visit, however, many of the automated segments had been removed.

Human factors considerations in design. Designing for safety and ergonomic integrity was stressed at the Clyde plant. It was the responsibility of the project engineer to make sure that the equipment was designed according to ergonomic principles and for safe operation. Design reviews were conducted with industrial engineering and safety officials to assure that nothing was overlooked. Both the design engineer and the safety engineer spoke of the need to have adequate safety reviews early in the design stage. Even though the design reviews were held before fabrication of the equipment, it was difficult to anticipate many of the problems. In spite of the periodic reviews, many modifications still had to be made after the equipment was built. Potential hazards could be more easily visualized at that time. Even though this caused some engineering inconvenience, all safety recommendations were thoroughly addressed.

In keeping with the "Japanese approach" to manufacturing, the assembly line was built on a nonsynchronous power-and-free principle. Between workstations the lines were powered, but at the workstations the lines were unpowered conveyors. This meant that movement on each segment of the assembly line was independent of the other segments. Each assembler was able to control the release of units onto the next powered line. Units were not to be released unless the quality was right. Minor perturbations in the overall line speed were not expected to seriously affect total line performance.

Problems encountered. There were many problems encountered in the start-up of Line 3. A number of assumptions inherent in the "Japanese approach" were overlooked in the design of this highly automated system. One was that all of the incoming parts and materials had to be virtually perfect. In manual assembly the assembler compensated for irregular parts, but automation was not quite so forgiving. Part tolerances had to be held within very tight limits, and many more attributes had to be specified. This meant that new relationships had to be established with vendors. A second assumption was that assemblers would use the line-stop buttons only for quality problems and not for other reasons. This meant a change in training and shop-floor culture. A third requirement was that the product itself had to be able to withstand different pressures and forces associated with automated handling. Robot applications were found to be too slow for high-speed assembly processes. Part location precision had to be held to tight tolerances. Many of the problems encountered with start-up were not directly associated with the equipment, but they highlighted many weaknesses in the total manufacturing environment.

The difficulties with the start-up of Line 3 affected the assemblers. Many people did not wish to be assigned to the line, which was often down. Assemblers would be sent home if the problems could not be corrected within a reasonable time. It was frustrating for the people to come to work expecting to do a good day's work only to find many difficulties beyond their control. One individual remarked that the old model washer lines would be working with cheerful, happy people, but on Line 3 there were only grim faces.

Evaluation of success. As a project, Line 3 came on line on time and within budget. It was capable of making the target production from the very beginning. Product quality levels were improved. The line did have the flexibility to produce all of the product models in the LEAP line. On the negative side, line downtime was very high, and scrap and rework levels were high. The operator learning curve for assembly skills was about one month, but the line required several years of modification before it met overall expectations.

At the time of the case study visit, the line was working well. It ran all of the low-volume production runs. It met production rates and quality exceeded expectations. The overall perception of the success of the line was that it came up to speed far too slowly and with too much anguish. It was important to note, however, that even during the most difficult times, the line far exceeded what the company's competitors had judged to be maximum assembly line output. Quality levels were also very high in comparison to earlier models.

Lessons learned. A number of valuable lessons were learned in the implementation of Line 3. The necessity of having consistently high-quality raw materials and parts for highly automated systems was made very clear by the implementation of this line. The necessity of having both assembler and skilled trades training was also apparent. Another point was that the design of the line with its widely separated workstations prevented the formation of social groups on the line. The lack of informal work groups was a negative factor for the assemblers. It affected their ability to cross-train each other on other assembly tasks, and this prevented assemblers from rotating jobs to alleviate boredom.

The advantages and usefulness of informal work groups had been overlooked in the design of Line 3. Voluntary job rotation within a classification was permitted at the Clyde plant. Assemblers formed informal groups, trained each other on adjacent operations, and then rotated among the workstations at will. Social groups formed and people migrated to the type of work they preferred. Some stayed in small groups where conversation was possible; others took up the "loner" jobs. There was apparently enough variety of work situations that everyone could find something to his or her liking. In a "family" factory where the employment level was two-thirds of the town's population level, this type of social interaction was extremely important. The distance between workstations on Line 3 prevented this type of work organization from happening.

Whirlpool Assembly Line 4

Project participants. The project engineer for Line 3 also designed Line 4. For Line 4, however, the approach to design was radically different. The engineer drafted three alternative plans for the line. Then, at an all-day meeting with over 100 assemblers, maintenance engineers, skilled trades people, and the full management staff, the three alternative plans were reviewed. Line 4 was essentially designed at that meeting. The best features of all three proposed designs were incorporated. Much to the surprise of the design engineers and others in management, a suggestion to eliminate the stop buttons and individual control of the line came from the assemblers. The assemblers preferred the discipline of a structured pace on the line to the unanticipated interruptions caused by others stopping the line. After the all-day meeting there was little continuing involvement in the actual equipment design by these participants. The design engineer implemented the plan according to the consensus reached in that all-day group meeting.

On this line the product designers from the corporate group were not involved to a great extent in the early stages of the assembly line design but did become involved in later stages as efforts were made to improve the manufacturability of the product. The safety engineer was also involved in the design reviews and equipment tests.

Design and installation. The vendor who built Line 3 also built Line 4. The same good relationship that had been experienced with the previous project also prevailed on this one.

The level of automation on Line 4 was significantly less than on Line 3. The corporate officers reversed their request to make the lines totally flexible to accommodate all models. Line 4 was designed to run the 27-inch and 24-inch models only. This permitted a less complicated line design. Although the number of automated stations was smaller on this line, the design engineer made sure that the line could accept automation at a future time if so desired. He described it as building the foundation of a house. As long as the foundations were solid, additions could be made without difficulty.

The line was power-train driven, and the line speed was under automatic control. Groups of four workstations were combined into zones. Between each of the zones was a buffer of three product units. Workers in a zone could stop the line within its own boundaries in order to resolve a problem. The three buffer units downstream from the zone would proceed to the next zone so that it could continue working without interruption. Up to six units could accumulate in the upstream buffer so that the preceding zone could continue work also. As soon as the limit of six was reached the line would restart at an accelerated rate in the zone that had been stopped. When the line had stabilized with three units in both the upstream and downstream buffers, the line speed in the affected zone would automatically resume a normal rate. This gave the work zone operators some control over their immediate workstation without affecting the entire line. It was necessary, however, to make up lost time after the line started again.

Human factors considerations in design. The assemblers on Line 4 had requested the removal of the stop buttons that allowed individual control of the line. That feature had caused too many disruptions on Line 3. Disruptions because of parts, material, equipment or assembly problems had created more frustrations for the assemblers than the discipline of a steadily paced but problem-free work routine. The other feature of Line 4 that was designed with human factors in mind was the grouping of workstations in close proximity to each other. This allowed for increased social interaction among the assemblers, a highly valued element of their work experience.

Safety and ergonomic considerations were well executed in this line, as they had been throughout the plant.

The benefits of having assembler participation in the design appeared to have a positive effect on the acceptance of the line and of assembler loyalty to the line. It was highly rated in an assembler interview.

Problems encountered. Beyond normal start-up adjustments no one spoke of any particular difficulties with Line 4. The effectiveness of start-up was so superior to that of Line 3 that all of the comments were very positive.

Evaluation of success. Line 4 was implemented on schedule and within budget. Quality levels were excellent. The striking success of the line was demonstrated by achieving production targets consistently after just 10 to 12 weeks of operation.

Line 4 was a success according to everyone who discussed it. The learning curve was a matter of a few weeks. Assemblers were pleased to see their suggestions implemented. Engineering and management were extremely pleased with the results. It was a smooth running operation from the onset.

Lessons learned. Training for operators and for skilled trades had not been covered as well as it might have been. There had been somewhat limited opportunities for the skilled trades to visit the vendor site where much advance training might have occurred. Training for assemblers was on the job and minimal.The question of how to implement training effectively for 150 assemblers appeared to be a significant factor in the failure to do so.

Whirlpool Assembly Line 2

Assembly Line 2 was designed by another engineer who followed the same approach to design that had been followed on Line 4. Participation in the design of this line came from supervisors, key hourly personnel, superassemblers (multiskilled assemblers), and utility people. The project engineer took the responsibility for gathering the input from these different groups.

This line had even less automation than Line 4. For the most part it was a retrofit of an older assembly line. This line used an index-and-dwell operation where the line would advance assembly units to the successive workstations and allow just enough time to complete the required operations before advancing them to the next.

Because the line technology was familiar, and they had had experience with the new product models from the first two lines, the assemblers were very comfortable with Line 2. This line was fully operational within two weeks.

Whirlpool Assembly Line 1

Project participants. Line 1 was a pilot project and, at the time of the case study visit, the line had been in operation only about five months. In that short period of time, however, the results were gratifying to all those people who were involved in it. The objective for this line had been to build a high-commitment, self-managing work unit. Most attempts at the development of this type of work group had been at greenfield sites. The Clyde plant was one of the pioneers that had attempted to initiate the concept in an existing plant. This was also the Clyde plant's first attempt at using a team approach to the design of an assembly line.

The idea of developing a high-commitment work force originated with the top echelon of corporate staff. It was a top-down decision aimed at pursuing the competition. The idea of a high-commitment work force was new to Clyde management, but it made the decision to phase in the concept with the introduction of a new production line. The new line was to produce the stacked unit, Thin Twin washer, which would be shipped to a sister plant for installation of the dryer unit. The Clyde management staff was the "coordinating committee" with respect to this new line.

The coordinating committee chose the manager of assembly operations to be the chairman of the design team. Other members included the manager of involvement, the manager of human resources, and the process engineer. This group invited the plant's communications specialist to be a part of the team as well. The purpose of the design team was to "evaluate new manufacturing concepts and systems and introduce these systems of production in the High-C line."

Five assemblers were included on the design team. These individuals were chosen on merit through a rigorous screening process. An initial group of 60 assemblers applying for membership on the design team included those who had responded to an employee interest survey. The concept of the High-C line was reviewed with them at a luncheon meeting. Those who were interested then had to write letters explaining why they wanted to be on the team. Ten finalists were chosen on the basis of the letters. Each of these individuals was personally interviewed, and five were selected for the design team.

The role of the design team was clearly defined and some operating guidelines were provided. The role and guidelines are listed below:

Role of the design team

- Determines requirements based on organizational analysis.
- Identifies areas for change consistent with the high-commitment philosophy. [see Appendix at the end of this chpater.]
- Designates, supports and directs study teams. (Groups went out as study teams to do standards, studies, and support work.)
- Sets priorities in consultation with coordinating committee.
- Sanctions study teams' recommendations.
- Develops employee strategies.

- Keeps coordinating committee informed of progress.
- Communicates with total organization.
- Models behaviors consistent with Whirlpool philosophy.

Guidelines

- Utilize current policy guidelines. Follow standard company policy on seniority and bidding of jobs, etc., *but* be willing to look at nontraditional policy.
- Utilize current discipline policy.*
- Operate within budget constraints.
- No changes in wage structure (at least initially).* [All assemblers at the plant belonged to one classification. It was understood that the high commitment group would be multiskilled.]
- Revisions to benefit plan should not be included in evaluation.*
- Production start-up would be April 18, 1989.

A consultant was engaged to advise on the development of interpersonal skills and problem-solving techniques and to help set up programs. Twenty-five people from the Clyde administrative staff attended a two-day seminar on the high-commitment concept. After the 10 members of the design team had been chosen, they went through a three-day training period. Additional training and consultation were provided as the project proceeded.

Design and installation. This assembly line was designed to be a low-volume line that produced the washer component of a washer-dryer combination assembled at a sister plant. The washer component had no outer panels, but was built up on a shipping skid. The shipping frame required a new engineering design, but, other than that, the assembly line design was typical of low-volume assembly lines for this type of production.

What amounted to rather routine engineering work was a totally new experience for most of the design team. Its task was to design the basic configuration of the line. The team had to plan production elements, such as the flow of material into the line, labor requirements, and production standards. The study team did this with the resources of the plant at their disposal. Through this process of designing the production line, the assemblers began to see the problems from the perspective of management.

The team met regularly. It had formal meetings once a week in which the work of the study teams was reviewed, assignments for the following week were made, and agendas were set.

After the line was built, the original five assemblers had the opportunity to choose the 17 additional people who were to be members of the High-C work group.

*The three items identified with asterisks were not to be changed, otherwise the design team could break with tradition in any manner it considered appropriate.

Interviewing and selecting work group members was a new experience for them. From the reports of the assemblers, this was initially even more difficult than planning the production line. The outcome of this experience was that the assemblers had an even greater appreciation of the problems of management.

Of all of the LEAP assembly lines, Line 1 had the lowest level of automation. It was almost primitive compared to the others. During the case study visit, several individuals noted that there had been a steady decline in the degree of automation associated with each successive line.

Human factors considerations in design. The project engineer designed the mechanical equipment for Line 1. The plans were periodically reviewed with the project team, which approved the designs but provided very little input into the actual equipment design.

The design of the workstation layouts, however, was a different matter entirely. The five assemblers actively participated in the workstation layout. The project engineer had no specific preferences for the way that work would be accomplished on this line. To encourage and solicit the involvement of the assemblers, he had them work with him in his drafting room for two weeks. He did not lead them into preconceived solutions but insisted that they analyze the work required and design the workstations according to their preference. His role was to show the feasibility of their solutions. The project engineer was comfortable working with the assemblers on this basis, but it took the assemblers awhile to realize that they were being given the responsibility to make many of the decisions themselves.

Problems encountered. The problems encountered on Line 1 were primarily those of adjusting to new working relationships rather than equipment problems. The work group had been given training on problem-solving and interpersonal skills, but these had not always been easy to apply. It was still necessary to get production out, and the group often felt the pressures of this responsibility. There was not enough time to do both the problem solving and meet the production schedule. There was a lot of frustration and anxiety.

In an effort to relieve some of these pressures, management assigned a team coordinator to the work group. The work group was supposed to have been self-managing. This individual was immediately perceived as a supervisor, which made it appear that management was interfering with the work group's autonomy. This added one more problem that the work group had to work through. To the credit of the team coordinator, he did not step into the role of a supervisor but confined his activity to helping the work group resolve its manufacturing problems.

What the group had lacked by not having an experienced supervisor was the accumulated knowledge of ways to get something accomplished in a complex organization. The work group did not know how to tap the resources of the organization. The team coordinator was able to provide this support without interfering with the autonomy of the group. Regular support from the organizational consultant was also provided to the group to help with the learning process in self-management.

Evaluation of success. The due date and the budget for this project had both been met. The outstanding measure of the success of this line, however, was the quality of the product it produced. One of the management staff remarked that the quality level on this line had far exceeded anything anticipated. The work group managed the quality levels of its product. If there was some question about the quality of any unit, the work group would ask itself, "Would you buy this product?" If the answer was "no," the unit was not shipped but was assigned to rework that was done on the spot.

Line 1 regularly met its production targets. When production requirements were reduced, the work group was given the choice of reducing its size or of finding more work for itself. The work group decided that it wished to remain intact and brought some subassembly tasks into the work center to supplement the work.

Management was pleased with the outcomes of the High-C line. Based on the production levels, quality, and initiative taken by the assemblers, even at this early point in the pilot program, the perception was that this line was very successful. Plans were underway to increase the scope of the project and to begin implementing High-C work groups in other areas of the plant.

The assemblers on Line 1 were enthusiastic. When asked what made the difference on this assembly line, one replied, "It's the involvement in my job that makes a difference. When I worked on the other lines, I parked my brain in my locker on the way into work and picked it up on the way out. I didn't need it for that kind of work. On the High-C line I have a chance to use my mind every day."

The assemblers on the other assembly lines seemed to be taking a "wait and see" approach to the High-C line. At first they thought that the high-commitmentline people would be singled out for special treatment by management. They were a bit puzzled by the enthusiasm that the High-C work group had for its work, and its willingness to "go an extra mile" to get a job done.

Lessons learned. The lessons learned on the High-C line were that people are willing and able to contribute to the design of technical systems especially with respect to how the work was to be managed and the workstation designed. It was also learned that the autonomous work group will cycle through "ups" and "downs" as it learns how to handle its new responsibilities.

Management found that it had to learn a new style of interacting with the work group. Old management habits were dropped. Fewer authoritarian decisions were made. Managers and supervisors had to offer suggestions and facilitate rather than direct or dictate. Those who made this adjustment discovered that this approach to leadership style had a positive effect in dealing with the traditional work force as well as with the high-commitment line.

Summary for All Lines

The assembly line experiences at the Clyde plant seemed to bear out the theory that it is not always the best approach simply to throw a technical solution at a manufacturing problem, and that the involvement of people can facilitate manufacturing change. Even when there is nothing wrong with the equipment or technical concept, there are many other considerations, such as raw material quality, product design, and social forces of production that affect the total manufacturing environment. At Clyde it was found to be better to back off from a total automation concept and to work on the other elements of production. As the emphasis on assembly technology diminished, the emphasis on people involvement increased. The improvement was outstanding. In summarizing the total experience of the introduction of the LEAP lines, management noted that it took longer to get there but that production and quality had far exceeded their expectations.

The lessons learned through the implementation of the LEAP lines tended to be positive experiences on which the Clyde Division wanted to build in future endeavors and in the daily relationships with its customers, shareholders, employees, suppliers, and community. To provide direction for this effort, the Clyde Division put together a philosophy statement of its responsibility to all those groups that hold a stake in its success. The text of this statement can be found in the Appendix at the end of this chapter.

HUMAN FACTORS CONSIDERATIONS DURING SYSTEM DESIGN

As each line was introduced at Whirlpool's Clyde Division, more and more emphasis was placed on human factors issues. When the first line was implemented, automation had been thought to be the answer to finding the most efficient production. Although safety, health, and ergonomics were considered in the design of the line, other human factors, such as training and design participation, were, for the most part, ignored. By the time the second, and later, third lines were being conceptualized, Whirlpool had learned from their mistakes with the first line. In the design of these two lines, operator involvement had become very important. The newest line, the High-C line, involved even greater steps in the consideration of human factors. This line introduced many new and aggressive philosophies concerning operator involvement in design, control and modification of the line, and in training. Because of the evolutionary development of the lines, each of the human factors policy areas will be discussed in the order in which the lines emerged.

Assembler Control

The issue of assembler control had been an area that was looked at from the beginning of the first line. The first line was initially very unsuccessful, partly because of the issue of assembler control. The design of the assembly line provided stop buttons for almost all of the approximately 180 assemblers. The idea behind the buttons was that the assemblers would not let the product pass their station until they were satisfied with the quality. It was soon found that these stop buttons were used for many other reasons besides quality improvement. Assemblers were stopping the line to light up a cigarette, talk to someone, or just take a break.

The idea of the "people controlling the line" did not go over very well with the management, nor with many of the assemblers who became frustrated with the line's frequent stops. In lines 4 and 2, the stop buttons were placed at four logical areas along the line, so if there were a problem, then one designated person could stop the line. In these lines, "zones" were placed between the assemblers, allowing the assembler to fall behind for a time and then catch up quickly. The assemblers were then able to produce good quality, without worrying about clogging up the system. The first three lines successively had less and less assembler control as they were designed.

Assembler control was a definite part of the design of the High-C line. Each of the assemblers had control over the speed of the system and were able to stop the system at any time. The line was smaller, had fewer assemblers, and the workers were able to play a more important role in assembling the product.

Modification of the Operation

The first three lines, being strictly assembly line operations, did not provide much opportunity for the assemblers to modify the operation along the lines. Though the operation itself could not be modified, the assemblers did modify how they would accomplish their individual tasks. Along all three lines, many assemblers were using very crude, yet efficient procedures to help simplify the job at their workstation. Some would use tape on their hands and wrists to hold screws or bolts, while others would even build small pieces of hardware to help in the moving and placing of their particular part.

On the High-C line, assemblers were encouraged to make any modifications of the operation, as long as it was checked with the supervisor and discussed with the rest of the line members. If someone came up with a better way to run a particular part of the line and it kept the quality high and was cost effective, it was usually implemented very quickly. At one time during production, the line was asked to produce an extra 250 washers. The team discussed the problem and decided to take away some of the subassembly that they were doing and to modify the line to some extent. The 250 washers were produced without using any overtime.

Feedback

Assemblers were provided with feedback, either verbally or through the use of a daily newsletter. Bimonthly meetings, which the management called ABC meetings, were held by the upper management in the Clyde Division. During these meetings, quality, production, inventory, or any other issues important at the time would be discussed. This information would then be passed down to the engineers and managers. They would then pass the information down to the supervisors, who were then responsible for talking to the assemblers. The feedback was presented to the assemblers on a "need-to-know" basis. However, if the assembler wanted to know any further quality or production updates, he or she was able to ask the appropriate person to get that information. Besides the bimonthly information, at least once a week other important material from the management and engineers would be passed throughout the plant.

Every day, a newsletter was printed and placed in well-traveled places for anyone in the division to read. *The Link,* as it was called, provided information on what the competitors were doing, new philosophies and technologies being researched, along with quality announcements. The Clyde Division had a research team that was solely in charge of gathering such information.

Audits were done on the machines in the Customer Acceptance Lab, and this information was then passed down to the assemblers. The members on the High-C line received direct feedback from the Marion Division, where the assembly of their product was completed. Marion provided the assemblers with quality reviews and future production demands.

Design Involvement

Involvement by assemblers in design had truly evolved over the four lines. The assemblers on the first line, Line 3, had no say in the design of the system. Corporate management was interested in entering the automated style of assembly. The line was highly automated and furnished with the renowned stop buttons, which were discussed earlier. Looking back, division management later wished they had asked for assemblers' input on the first line.

As the concept of the second line, Line 4, began to develop, the management seemed to have learned from the first line. The assemblers became very much involved in the design of the second line. The division invited everyone, including all assemblers, to come to a breakfast on a Saturday to discuss the design of the upcoming line. The engineers introduced three proposals for the second line. Each proposal was explained, and the assemblers were asked for their opinions on each. The second line then was designed from portions of each proposal, with the help of information the assemblers provided. This type of participation was also used in the development of the third line, Line 2, but to a lesser extent. The second line was running up to full potential within 10 to 12 weeks after start-up and the third line within two weeks, compared to the three years it took for the first line. This was attributed to the assemblers' suggestions and overall involvement of the workers. By this time, the engineers said, they knew what the assemblers liked and disliked.

Involvement by the assemblers in the fourth line, Line 1, was unprecedented. After the five assemblers were picked for the fourth line design team, the team was almost totally responsible for the design of their smaller assembly line. The team was given a certain area in which they could work and told to come up with the assembly line. Chalk and tape was distributed giving the assemblers the opportunity to design the layout of the line. The sequencing, placement, and designing of the individual operations along the line were all in the hands of the assemblers. They soon found this responsibility to be overwhelming and had to seek the knowledge of the engineers. With the help of the engineers, the team was able to come up with a layout that they felt would work best from the standpoint of quality, overall efficiency, and cost effectiveness. Besides working on the actual physical lines, the five assemblers held interviews to pick an additional 17 assemblers for the line. By giving assemblers this degree of control over design of the line, division management instilled pride among the team members. The members felt an ownership to the line that they had helped create. As a result, the quality on this line had been excellent almost from the start.

Involvement in the design of the product had always been encouraged companywide, and was also perceived in this division. If anyone had a suggestion regarding the design of the washer, they could let that request be known by turning in a written statement of their idea.

Safety, Health, Comfort, Stress

Throughout the plant, it could be seen that safety was very much stressed. The building and equipment were designed according to OSHA standards. The equipment was issued with warnings along with descriptions of appropriate actions in case of emergency. Machines were well guarded, and there were rails along the platforms and walkways. Aisles were wide and kept clean and free of debris. Injuries were expected to continue to decrease to 13 percent of their 1983 starting level by the end of 1989. Assemblers were encouraged to modify their workstations to make them as comfortable as possible.

When production started on the first line, there was much stress, largely because the automation was new and the assemblers lacked training. It was about three years before this line ran as smoothly as management wanted. Throughout those three years, much frustration was found among all the people involved in the line.

The other lines had much less stress because of the lessons learned on the first line. Because of the zones in the lines, which by then were used in most areas, assemblers felt more at ease in their particular tasks. If they fell behind, they could always catch up within the next minute. On the first line, assembly stations had been placed quite far apart, discouraging interactions among the assemblers. This was disliked by the assemblers, so the work stations had been positioned closer together in assembly clusters in the next three lines. Assemblers were then able to associate with other assemblers, and a happier working environment evolved.

Maintenance

Assemblers were encouraged to fix jams along the line, but not to do any real maintenance. The Maintenance Department found that many times assemblers would try to repair a part of the process and then not report it to the supervisor or other authority. If, farther down the assembly line, the product ended up with a problem, the maintenance workers would have no record of what had caused the problem and what attempts had been made to correct it.

Because the assemblers on the High-C line had more information about their product and process than the assemblers on the other three lines, they were encouraged to do their own minor maintenance. Furthermore, the assemblers on all four lines were encouraged to help the supervisor and maintenance personnel to understand exactly what was not working and why.

Worker Participation After Installation

Postinstallation worker involvement was responsible for a great many of the major improvements that had occurred in the division. The assemblers, for the most part, did not have much to say about the installation of the first line, but they did get involved after it began production. At first the line had many problems, as previously mentioned. Very quickly, however, management, engineers, supervisors, and assemblers began having daily meetings on what could be done to improve the line. One of the most important changes that the assemblers wanted, and that management agreed with, was the removal of most of the stop buttons given to the assemblers. Other changes were gradually made until the line reached the production and quality levels needed.

After the second and third lines were installed, all assemblers were encouraged by management to suggest alternative ways to do particular tasks or to make better product. This approach was evident throughout the plant. In the cafeteria, a sign supported this by saying, "Cost reduction is everybody's job. You can suggest a better way. Turn in your idea now." Turning in ideas was promoted by giving a person a raffle ticket for each idea submitted.

When the fourth line was installed, worker involvement never stopped. Assemblers were totally in charge of the line and, therefore, were held responsible for performance of the line. When difficulties occurred on this line, the team assemblers would work together, using other resources if necessary, and solve the problem. They became involved in issues such as direct labor utilization, line balancing, scheduling, and other assembly line considerations. The assemblers on the High-C line were not able to "leave their brains at the door" when they came to work.

The assemblers had begun to think of running the High-C line as if it were "their own business." This philosophy seemed to be working, and was slowly being introduced to other areas in the plant, such as the machine shop.

Ergonomics

The Clyde Division sought to achieve good ergonomic design in all of their lines. Most of the material handling was done by conveyors, and heavy lifting did not seem to be an issue. Assemblers were encouraged to arrange their own workstations to make their jobs as physically easy as possible. Assemblers within cells were invited to switch jobs to avoid boredom and repetitive use of body movements.

Training and Skills

The division took some time to fully understand the importance of assembler training. As the first line was going into production, newly hired assemblers were essentially asked to come in and start right up on the line. The equipment had been coming in too fast, and training had become a problem. The new hires spent some time with experienced workers, but in less than two days they were working entirely on their own. Supervisors estimated that it took about a full month for the assemblers to be fully confident in their tasks. As a consequence, it took longer than expected to have the line up to full capacity.

When production started on the second line, the same scenario took place. The budget did not allow the assemblers to get any outside or extensive internal training. Some of the skilled people were taken to the equipment vendors, but the assemblers received their training on the job. The third line, being the simplest line and least automated, needed less training, but again, no training was provided before placing the assemblers on the line.

In line with the new High-C line philosophy, training was provided. When all the team members had been picked for this line, they spent three days at an off-site location in intense training. Time was spent on problem solving, communication skills, and the brand-new philosophy that was to be implemented on the line. After returning from training, the assemblers struggled for a while. More training was obviously still needed.

A training committee was formed from among the High-C team and placed in charge of the ongoing training that the assemblers would need. They developed a three-phase training program that included multiskill, indirect, and social skill training. The multiskill phase involved learning all the skills on the line, not just those found in a particular cell. The goal was for all the team members to become fully multiskilled within a nine-week period.

The second area in High-C assembler training included explanations of what was actually involved in other areas committed to the production of their product. The fabrication and paint areas were discussed, along with the roles that engineers, vendors, and scheduling personnel played in the process. Through this type of training, the members of the line obtained a broader view of washer manufacturing.

The last area of training consisted of further communication and people skills, such as negotiating conflicts. This three-phased training program, not totally prescribed at the start of development of the High-C line, had become formalized within the division.

Summary

Safety, health, and ergonomics had been considered in the design of all four lines and had been well implemented.

When the first line was being built, it was felt that the role of the assembler had been carefully considered, but management failed to ask for the assemblers' opinions. The division struggled during the first few years of Line 3, and by the time Lines 4 and 2 were installed, the importance of assembler involvement was apparent. The High-C line truly illustrated total worker participation. Throughout the evolution of each line, consideration of human factors seemed to become more and more apparent.

At the time of the visit, the desire to involve everyone in the division in improving quality and staying cost effective was very visible. The division had learned from its experiences and was now committed to a new approach for the coming decade.

IMPACTS OF THE LINE DESIGNS

The LEAP project made a significant impact on the way of doing business at Whirlpool. The operation had just experienced a long transition period of intense concentration on the implementation of the four new production lines. Although the implementation period had taken longer than expected, the increased capacity and outstanding quality levels provided a return on the investment that exceeded the most optimistic expectations. As the attention of the plant turned from process implementation to one of continuous improvement, issues of quality and changing the operating culture of the plant became the focus of attention. The LEAP project had made a significant impact on how these issues would be handled by both management and operating personnel.

Corporate and Plant Management

Corporate management was pleased with the results at Clyde. A major financial investment had been made at that location, and it had far exceeded the expected return. Corporate management recognized the achievements brought about by the LEAP lines. It responded by reducing its close scrutiny of the plant and by granting greater autonomy to the local management. With the operation at Clyde well under control, corporate focus was able to turn to other issues.

With the increased autonomy for Clyde management came added responsibility and accountability. Success brought new challenges for even greater success. The plant management had to expand its horizons. Rather than look inward to the problems of production, it looked outward to its surrounding environment. There was more awareness of the highly competitive market and the need to respond quickly to moving targets. The plant directors and management changed their leadership and management styles. They engaged in more joint problem-solving activities. There was more emphasis on planning and strategy, and fewer crisis-oriented sessions. Personal productivity increased. People were "wearing more hats" as their responsibility broadened.

As emphasis shifted from the development of new production lines to the improvement of the existing lines, product quality became a central focus of plant management. Standards of acceptance were higher. There was a new willingness to reach for these standards. Production concepts such as "zero defects" had become realistic production goals. Previously, the notion of zero defects would have been perceived as a flight of fancy.The experience gained in the LEAP introduction gave management the confidence it needed to strive for these new targets.

Design Team Engineers

Prior to the large-scale projects on the LEAP lines, manufacturing engineers had individual responsibility for specific equipment projects. Over the course of the development of the LEAP lines, design became more of a team effort. Design reviews, which involved different people at different organizational levels, were initiated. These reviews became a forum in which anyone could voice an opinion and be heard. Ideas were generated into solutions. This approach to equipment design began informally but later developed into a series of more formal design review meetings. As an outcome, there was more cooperation among the various functions.

Training

The introduction of the LEAP lines brought much more emphasis on training, especially for mechanics and craftsmen. It was the responsibility of the manufacturing engineers to see to it that this training occurred and to find the people to do it. Some of the training was done by maintenance engineering; other training was done by vendors.

Training for operators on the assembly lines had previously been limited to on-the-job training. Much of this had been done within the informal work groups. With the High-C line, however, there was a need for more extensive training. Operators on this line were responsible for more tasks at each workstation. Rotation within this group meant that the individuals would eventually have to learn how to assemble and line-test an entire washer unit. Trouble-shooting quality problems required the ability to do minor rework and repair on units before they ever left the line. In addition to skills training, the High-C line assemblers had to have increased interpersonal and problem-solving ability. They also had to learn how to tap the resources of the organization. The responsibility for this training came from the supervisors who took on a different role of leading and facilitating activities rather than "bossing" the assemblers. The High-C work concept was scheduled for implementation in additional groups, so training activity was expected to remain high.

Supervisors

First-line supervisors found their jobs changing, and they were under much more pressure. They had become the primary information channel between the organization and the work force. Assemblers were asking more questions, and supervisors had to explain why certain changes were necessary. They had to explain the reasons for engineering process changes. The newer equipment was more complicated, however, so supervisors were forced into greater dependency on support people. On earlier lines supervisors had been more involved with machines; on the newer lines they were more involved with people.

Operators

As each of the four LEAP lines had been implemented there had been more involvement of the assemblers with the design of the process. Engineers were beginning to seek the input of the assemblers. The assemblers were given opportunities to sit with the engineers during the project conception and to inspect the equipment during its design and fabrication. Input from the assemblers on product manufacturability was also encouraged. A corporate program to elicit recommendations for product and process improvement from the operating personnel resulted in over 3,300 individual suggestions from the Clyde plant alone. The operators on the High-C line were enthusiastic about the opportunity to become totally involved with their work.

Maintenance

Maintenance was one of the skilled areas at Clyde. Once new equipment was installed, the maintenance people had the ongoing responsibility for it. To reduce later complaints about the equipment, maintenance people were given the opportunity to check and revise drawings, and to participate in the planning of installation and debugging. At the production level, barriers among the crafts started to disappear.

Long-Term Impacts

The local management was willing to predict what some of the long-term impacts of LEAP lines might be on the Clyde operation. These predictions covered the people, the product, the process, and the technology.

From the standpoint of people, Clyde management expected the shop-floor culture to change. There was a plan in place to continue development of a high-commitment work force for the entire plant. It was expected that this would also have an impact on machine design. Decision making would be pushed to the lowest organizational level possible. Work groups would assume many of the responsibilities of the supervisors.

A need for more technical people between the craftsman level and the engineering level was anticipated. Training was expected to be given a greater emphasis. More training on computers, diagnostics, and other technologies would be required. Rotation among positions would be delayed until the replacement was fully trained.

The financial rewards to people at various levels were seen as narrowing. There would be less difference between exempt and nonexempt compensation. The work force would share financially in the good and bad times for the company.

There would be a greater emphasis on product concerns. Workers would be given greater opportunity to suggest how the product could be improved. It was anticipated that they would continue to ask challenging questions on strategic issues such as the expatriation of production operations that could be performed equally well at the Clyde Division.

Trying to save money on the front end of projects would be discouraged. There would be an emphasis on quality and "making it right the first time."

There would be a change in production philosophy. Process issues would be primarily those dealing with the ability to respond quickly to change. The use of direct-flow processing and just-in-time inventory principles were expected to increase. Products would be run as needed rather than building banks of inventory. This would place even more emphasis on the ability to respond quickly and decisively.

Computer-based information systems that could report the status of production and equipment would be required. The ability to diagnose both production and quality problems at the source was seen as crucial. The assemblers had already been asking for computer-linked information on the production requirements of their customers so they could respond to downstream demands. This trend was expected to continue. All employees would expect more information on shipments and quality.

Advanced technology was seen as a superior way in which to control the process. New methods of control were developing rapidly and were increasingly sophisticated. Enhanced information systems would be required to support process control. There would be a need for people to keep up to date on these developments. It was expected that there would be continuing support from management to provide the learning opportunities.

SUMMARY

The experience on the LEAP lines appeared to build confidence in people. They had proven to corporate management, and themselves, that they could successfully implement a major change such as this and were ready to take on increasing responsibility. They described this as the ability to raise the goal while in flight.

APPENDIX FOR CHAPTER 9

Clyde Division Philosophy Statement*

We believe that the Clyde Division of Whirlpool Corporation has a responsibility to provide for and balance the interests of its customers, its shareholders, its employees, its suppliers, and the community in which it operates. We will provide an environment for change that is built on a foundation of mutual trust, where each person is treated with respect and offered an opportunity to participate in this change. We will relentlessly pursue quality and cost improvements to be the world leader in quality and cost for our industry.

Achieving the most effective overall results will require open and meaningful

*Source: Whirlpool company document.

communication that encourages initiative, experimentation, and the generation of new ideas. Each of us must recognize and abide by the following:

1. *TRUST*
   Everyone will be treated with fairness, respect, and dignity.
2. *CUSTOMER SATISFACTION*
   Each individual's principal goal is customer satisfaction. (Your customer is the next person to receive the results of your efforts.)
3. *TRAINING*
   People throughout the organization are capable of taking on more responsibiltiy and making a more significant contribution with appropriate training and opportunity. Each individual has the responsibility to improve their job skills.
4. *HOUSEKEEPING*
   Each person takes the responsibility and pride in maintaining a clean, safe environment (a place for everything and everything in its place).
5. *TEAM WORK*
   Each individual accepts ownership for the results of their work and is committed to cooperating with other members of the team to assure the success of our company.
6. *DECISION MAKING*
   Our organization will be most effective when people closest to the work have the knowledge, skills, and authority to make necessary decisions and do so.
7. *ACCOUNTABILITY*
   Each individual will measure their actions against this philosophy.

10

ANALYSIS OF CASE STUDIES

The five case studies present an interesting blend of diversity and similarity. The product lines of the five companies, for example, are quite different, but there are similarities of approaches to new technology in a number of instances. There are differences as well, and it will be useful to compare and contrast the information provided in the studies to see more clearly the human factors principles that should apply to the design and selection of advanced technologies.

Analysis of the case studies is organized as follows: Because of the high degree of commonality among the firms regarding health, safety, stress, comfort, and ergonomics, we address the areas of physiological needs as a group. We treat each of the other human factors policy areas separately. We then examine the case studies to see if they reveal other policy areas not identified by the survey. Finally, we examine the outcomes of the projects in terms of operating results. Using the insights we gain from this analysis, we present in Chapter 11 the principles distilled from this study.

PHYSIOLOGICAL NEEDS

All five of the companies studied are very large and are leaders in their respective fields. It is not surprising, then, that all had explicit policies regarding health and safety that were at least as stringent as those required by OSHA. Legal liability and public relations reasons combine with concern for employee welfare to make this area the one that has had the most complete, formal policy response.

In most of the cases a safety engineer reviewed and approved proposed configurations of new systems before they were built or bought. It was also common practice for the safety engineer to review a system once it had been installed. Many of the firms had gone so far as to publish manuals to guide engineers relative to health, safety, and ergonomics.

In matters less central to OSHA mandates, the companies' policies were much less concrete, as, for instance, in the areas of comfort and stress. Equipment engineers tended to show less concern in these areas.

None of the cases describe any participation of operators or mechanics in health or safety issues of new technology. There seems to be an implicit understanding that safety and health concerns must be handled by experts, that is, design engineers or safety engineers. This reliance on experts can produce systems that are theoretically safe, but practically dangerous. Even where there are explicit policies and expert attention, several of the cases demonstrate that it is difficult to anticipate all the problems that will arise. This is illustrated by the situation at the viewfinder assembly machine at Polaroid, where the combination of frequent machine jams and pressure to meet production schedules prompted operators to block out safety interlocks so they could keep machines going.

Stress also shows up during machine start-up and debugging. The frequent stoppages of the first Whirlpool line, for example, proved stressful and frustrating to the assemblers for many of the same reasons that were present in the Polaroid situation.

In several of the cases, engineers pointed out to the case writers that, because they were purchasing standard machines, they were relatively powerless to impose stringent rules for physiological factors on their suppliers. If a machine manufacturer fails to consider the operator's needs when designing a workstation, the customer frequently is unable to do anything about it until after the machine is delivered. Sikorsky, for example, had to build a platform around its Mazak machine once it was installed so the operator could work safely and comfortably.

EMPLOYEE INVOLVEMENT IN DESIGN AND SELECTION

Of all the human factors principles related to the introduction of advanced manufacturing technology, the one that appears to be making the greatest gains in recent years is that of involving operators, mechanics, and foremen in the planning stages. By getting the production worker involved, the designer exploits the user's expertise, and may gain the user's support.

Where our cases reported on sequences of new technologies being introduced in the same location, as in the Whirlpool case, experience led to recognition of the benefits of employee participation. New policies requiring employee involvement resulted. In the Whirlpool situation, where four separate assembly lines were designed and installed, the degree of employee participation went from zero for the first line to continuous involvement in the design team for the fourth. Operators eventually were even taking part in the decisions of the team.

The Sikorsky case described an implicit policy of employee involvement that included review of layouts, evaluation of proposed equipment and vendors, and preparation of a report of operator and foreman recommendations. Timken's Faircrest plant was a greenfield site, and there were no operators on the payroll to consult until after the plant was built and equipped. Maintenance personnel who were hired in advance of start-up did have a chance to test control station mock-ups and to check out equipment at suppliers.

Where there was no employee input during process design or selection, as was the situation in the Polaroid and Westinghouse projects, it was evident that operators could have made contributions to the design. This feeling was actually expressed by Westinghouse operators to the case writers. In the Polaroid robotic assembly line situation, there were a number of instances of poor location of machine components or feeding mechanisms that would have benefited from operator insights.

Although operators and mechanics are unlikely to have global perspectives regarding system design, their daily contact with the machines and software make them experts on what works and what is likely to fail. Examples of operator input cited in the cases included better equipment placement, consolidation of work, and changes in operations sequences to avoid product damage—all instances in which familiarity of the operator with the process and product resulted in improvements in operating efficiency or product quality.

Equally important are the benefits to be derived from improved employee attitudes toward new technology because people were involved in its design and selection. New technology frequently is considered a threat by employees concerned about changes in routines, skills requirements, or job security. As was demonstrated in the Sikorsky case, involvement of operators mitigated reactions toward a new system that was about nine times more productive than the machines it replaced. The operators felt they were responsible for the success of the system and felt a greater sense of involvement and pride in the process.

Formal programs of employee involvement in process design and selection clearly have costs. Not only are there costs in terms of wages paid to employees for "nonproductive time," but their involvement may also lengthen the planning process. We have seen in this study, however, that these costs can be more than offset by reductions in problems during start-up, by higher system efficiencies, and by better-trained, more responsive work forces. In fact, a substantial part of the cost of employee involvement during equipment design probably should be charged to training, as an essential feature of the installation of complex technology.

As the cases indicate, involvement in the design or selection of manufacturing systems can take a number of different forms. The most modest of these are information sessions or presentations made by designers that communicate design intent to employees—a one-way process. Greater involvement can be found in situations where there are periodic working sessions between designers and production workers in which viewpoints and suggestions are shared. More complete involvement is found when operating employees are represented full-time on the design and selection team, participating in the evaluation of designs and suppliers and in making decisions regarding the system. It would appear, from the five cases, that the more complete the involvement of operating personnel, the greater the chances of success for the new technology, both in smoothness of start-up and in operating performance.

CONTROL OF THE PROCESS

In contrast to the fairly well-defined movement toward employee involvement, the question of employee control of the process remains an issue among designers and managers. Senior technical people give strong arguments for denying the operator any control over process settings. In some instances, company representatives answering our survey told us that operators lacked the requisite skills to make control decisions. Others stated that the speed, precision, or complexity of the process was such that it was impossible for a person to control it. Our survey also revealed, however, that sometimes these reasons were underlain by a management philosophy that wished to use new technology to reduce or eliminate the influence of labor on the performance of the company.

In our five cases, policies relative to operator control of the process varied significantly. We even saw variance from one process to another within the same plant. In Sikorsky's spindle/cuff machining cell, for example, operators were permitted to edit programs only to adjust machine settings for differences in raw part size or composition. They could not change tool paths or sequences. Yet, in an NC lathe cell in the same plant, operators were encouraged to program all steps in the processing of parts on their machines. Professional programmers were instructing the operators in this cell in the best programming methods.

Westinghouse had an explicit policy of restricting operator control to clearly defined parameters. Operators in the FMS facility were able to enter tool offset corrections only if piece-part measurements indicated a need for change. Other opportunities for modification of the process appeared to be quite circumscribed.

We also saw variance between stated policy and actual practice. Polaroid, for example, had a policy of not allowing operators to use tools for equipment or process adjustments. The assembly machine design locked out the operators from intervening in machine operation. When problems arose after the machine had been put into operation, however, operator intervention became necessary. The design of the machine then made such intervention awkward, possibly dangerous. Operators were even given tools to do minor adjustments and to perform maintenance tasks.

Timken, on the other hand, had designed its steelmaking process with the intent of giving the operator control if and when needed. Information was provided promptly by computer to assist operator decisions, and all equipment had manual overrides so the process could be operated without benefit of computer systems when necessary. In practice, however, some of the automatic control systems worked so well that there was little need for operator intervention. The direction of design policy was clearly toward employee empowerment, diametrically opposite to policies that lock the operator out.

It is interesting that, in the series of line developments in the Whirlpool case, increasing participation of operators in system design did not necessarily result in increased control of the process by individual operators. In fact, the trend in the first three lines was to go from individual worker control over shutting down the line to zone control for which designated operators were responsible. Operator frustration over unexpected stoppages was given as one reason for this shift. In the fourth (high-commitment) line, individual worker control of line operation and speed was restored. Because a substantial amount of team building took place during the development of this fourth line, it is likely that stoppages, when they occurred, would have been accepted by the team members on that line as reasonable and necessary.

The divergence of policies relative to employee control and modification of a process appears to stem from two basic roots—management philosophy and technical imperatives. While it may be true that certain process steps are beyond the capabilities of humans to control, this need not be the basis for denying the operator any opportunity to control or modify other aspects of the process. The Timken policy of moving decision making and control down to the operator and of providing computer assistance to enable the operator to act, demonstrates how a management policy favoring operator control can be implemented even where process errors can be costly or dangerous.

FEEDBACK OF PRODUCTION INFORMATION

If an operator or mechanic is to be fully equipped to deal with problems that crop up at his or her workstation, timely information on how well the process is working must be fed back in sufficient detail to be useful. Information feedback is not only related to operator control and to operator modification of the process, it is also related to employee motivation and morale.

In accord with its design intent of giving maximum control to the operators in the steelmaking process, Timken's designers provided means for fast feedback to operators via computer. Metallurgical assays were quickly transmitted to the ladle-refining operators. Computer assistance in decision making was available. Equipment performance data were provided to the operators. Production schedules and operating results were available to and used by all operators. Thus, they had both workstation and plant performance data available to them.

The Westinghouse FMS system provided machine operation instructions (in machine code) and statistical process control charts to the operator for adjustment of the process. Production output and quality charts for the FMS were displayed in the area. The Sikorsky FMS cell also provided similar information and gave the operators the ability to call up on the computer screen the work queue scheduled ahead for their workstations.

In the other three cases, most of the feedback was accomplished by posting production data periodically in the work area, by distributing newsletters, or by word-of-mouth through meetings. In the Polaroid situation, one of the design principles enumerated by the machine designers was that the machine should tell the operator why there was a malfunction. It was not evident that the viewfinder assembly machine design actually incorporated this principle. Operators received little or no diagnostic feedback from the computer system.

Advanced manufacturing systems make great use of computers to control the system, to process information, to provide diagnostic and decision-making assistance, to disseminate information, and to provide an interface between the human being and the machine. It was evident from the cases that operators quickly learned how to use computers to manage their jobs and to keep informed. The enormous capability of the computer to provide readily accessible and prompt feedback is well established. The degree of feedback to employees, however, depends on system design policy. If full effectiveness of operator, mechanic, and foreman capabilities is desired, feedback to these people becomes an essential principle for system design.

Manufacturing systems can even be designed to enable an operator to generate his or her own feedback. Statistical Process Control techniques for recording and analysis of measurements by computer can be put into the hands of the operator, and the results can guide operator decisions regarding system performance and system adjustments.

MODIFICATION OF PROCESS

The authority to make more or less permanent changes in a process differs from the need to have a person make continuous or repeated adjustments to keep a process within control limits. As reported in Chapter 2, a number of respondents to our survey indicated that they did not want to have operators modifying the process. The implication in each case was that the operator lacked the skills or could not be trusted to make changes.

Sikorsky's case was a bit different. The reason for denying operators latitude in process changes was the critical nature of the helicopter parts the cell was making. Once the process steps had been proven, they were frozen by government procurement regulations as well as by management policy.

Although the Westinghouse case states that there was an implicit policy to design systems so operators might modify them within prescribed limits, there did not appear to be much opportunity for them to do this on the FMS. As was the case at Sikorsky, modifications were limited to adjusting tool offsets to account for differences in raw part dimensions.

Polaroid operators had little opportunity to modify the process. This was also true for the operators on the first three of the lines studied at Whirlpool. In the case of the fourth line, assemblers were encouraged to make modifications. When there was consensus among the team members, a change was made quickly.

Although we found relatively few examples in our study where operators were encouraged to modify the process, these tend to indicate that, where circumstances permit, it is a useful policy. By implication, if an operator is expected to modify process steps, he or she must have a sufficiently good understanding of the process and its effect on the materials being processed that changes can be made correctly and efficiently. Both training and trust are requisites.

SKILLS UTILIZATION

A few of the cases gave some indication that the machine system designers were encouraged to provide work environments that would fully utilize worker skills. The design approach at Timken's Faircrest plant was unusual in the degree to which the operator's abilities were to be utilized and even enhanced through computer support. The design of Whirlpool's fourth assembly line also took into account the ability of the operator team to manage the system for best results. Sikorsky's cell design profited from the skills already inherent in the skilled operators who assisted the designers, and there was significant encouragement from management for operators to keep developing their skills. In each of these cases it is highly likely that the operators, if asked whether their jobs supplied opportunities to use their abilities, would have answered in the affirmative.

The concept of multiskilling had been adopted in many of the situations (Timken, Sikorsky, Westinghouse, and Whirlpool). By reducing the number of job classifications at the same time they were cross-training people, these companies were opening up more degrees of freedom for engineers designing machine–human systems. The design teams at Timken and Whirlpool, in particular, took advantage of this opportunity.

In two of the cases, reduction of worker skill requirement or elimination of worker involvement altogether was a goal. This was true in the Polaroid case and in the first of the Whirlpool lines.

Most large companies have some form of internal job training for employees new in their positions. These firms also have programs for paying the costs of job-related education or training outside the company. The five firms in our case studies had such programs. The differences we found concerning whether or not skills development was actually being pursued appeared to be a direct consequence of the social and technical environment in which people worked. The technical challenges of the jobs at Sikorsky and the opportunity to use acquired skills ("trigging it out") encouraged the operators to seek both inside training from programmers and outside courses at a junior college. The breadth of competence expected of the steelmaking operators at Timken also led to extensive training and skills development. Westinghouse, on the other hand, had a new compensation plan that encouraged further technical skills development, but rules in the workplace limited the extent to which operators could exercise these skills.

By the time they were ready to design the fourth assembly line at Whirlpool, training of operators had expanded to include problem solving, decision making, interviewing, quality control, and use of company resources. Such a program is a far cry from an approach that assumes that workstations will continue to be manned by people with low skills, content with low pay and little opportunity for personal growth. Although Polaroid has a national reputation for its progressive attitudes toward human resource development, the Polaroid design team clearly considered the operator jobs on the viewfinder assembly line to be of this latter type. The resulting technical work environment was devoid of any encouragement toward skill development and use.

"Use it or lose it" seems to be a reasonable interpretation of what we found regarding skills. If the work environment does not expect or require skills to be used, there is little incentive for people to learn new skills. The work environment, shaped by technology and by company policy, is a major determinant of the extent to which human capabilities are used and expanded on the job.

MAINTENANCE RESPONSIBILITY

Responsibility for keeping the process in operating condition has traditionally been considered work outside the domain of the production operator. Typically, a firm has had maintenance mechanics who handle all maintenance and repairs, from emergency trouble-shooting and routine preventive maintenance to major machine overhaul or repair. As indicated in several of the cases and in responses to our survey, there is a trend toward assigning at least first-level maintenance duties to machine operators. In so doing, firms have also designed the process to make it easier for operators to diagnose trouble and correct it.

The increased interdependence of computer-controlled systems argues for very prompt response to maintenance needs. If one section of a system goes down for any reason, that stoppage is likely to affect other sections rather quickly. Long delays that occur while people and machines wait for a maintenance mechanic to appear on the scene become unacceptably expensive in lost productive time. When that problem can be readily diagnosed and corrected with simple tools by a knowledgeable operator, idle time is minimized.

Timken's Fairfield Plant expected operators to use the computer system to obtain first-level diagnosis of equipment problems. Timken's policy explicitly required operators to perform first-level maintenance. The equipment was designed to facilitate this. The Steel Business Unit production team operated on the principle that all members of the team were to assist in returning the facility to operation when there was a breakdown.

Minor first-level maintenance was actually being performed by operators on Polaroid's viewfinder assembly line, even though the line had not been designed for this and the operators had not been involved in the equipment design phase. As a consequence, some incipient maintenance and housekeeping problems created by the design were not detected until too late—when the equipment was installed and operating.

The more traditional separation of operating and maintenance roles was found in the other three cases. Sikorsky, Westinghouse, and Whirlpool had separate maintenance departments staffed with skilled tradespeople. There is evidence that long-standing jurisdictional agreements in union contracts have tended to block management efforts to give operators more responsibility for maintenance. Westinghouse managers, for example, told us they wanted first-level maintenance to be included in the expanded work roles of the operators, but that operators had not yet accepted this responsibility.

Operators who handle maintenance responsibilities require know-how in the functions of the system, in the use of diagnostics, and in the use of tools. Companies that adopt a policy of operator-initiated maintenance, therefore, will find that they are also committed to a continuing program of training. Rewards from such training come not only from improved system reliability, but also from suggestions for system performance improvements derived from operator understanding of how the system works.

PARTICIPATION IN POSTINSTALLATION DESIGN CHANGE

A decision to involve employees in suggesting design changes to equipment once it is installed implies more than simply activating (or reactivating) an employee suggestion system. As was seen in several of the cases, employee involvement at this stage is the result of employee empowerment through education and training, and of employee motivation by a responsive management that rewards them for getting involved.

Employee response, in terms of suggestions to improve the process, appears to flow most naturally when people are already significantly involved in system control and modification. This certainly was true at Timken, where operators felt a sense of ownership in the process. They were encouraged to recommend changes and were rewarded through a gain-sharing plan.

Active employee participation was also seen at Sikorsky, where the vice-president in charge of machining operations made a personal effort to see that engineers and managers responded promptly to ideas. Engineers learned to follow up on written suggestions they had received by visiting the person making the suggestion to clarify his or her intent. This practice produced dramatically higher rates of acceptance of suggestions. Cash awards provided added inducements to the employees.

Whirlpool also had increased the vitality of its employee suggestions regarding product and process improvement. Westinghouse encouraged operator suggestions, but there were indications that some supervisors were uncomfortable in coordinating, facilitating, or problem-solving roles, and this discomfort undoubtedly served to discourage suggestions from operators. In the Polaroid case, operators were clearly not expected to make suggestions for process improvement, and no formal mechanism seemed to exist to channel recommendations to people who could respond to them.

Only infrequently does a company take a giant step forward in process capacity, productivity, or quality. Most of the gains in manufacturing capability come incrementally, the result of continuous efforts at improvement. These incremental gains can, over time, produce enormous advances. Continuous improvement programs that make it possible and rewarding for employees to participate appear to be most effective if they are started when a new system is being designed and then continued through the life of the installed system.

OTHER POLICY AREAS

We have searched the cases and the survey responses for other possible policy areas that relate to the design and selection of new equipment. We find that the initial list is essentially complete. The only area that might be added as an engineering consideration is that of housekeeping. Many production processes are inherently messy—it takes skill and forethought to design systems that eliminate or manage all of the waste that is generated. Poor design can accentuate housekeeping problems. Poor training, motivation, and discipline of production workers can further aggravate these problems. Where we found high levels of operator involvement in the design and implementation process, we found clean, uncluttered work areas; where operators were more isolated from the process, the work areas tended to accumulate waste and dirt.

Housekeeping is both a system design issue and a human issue. It affects product quality, machine performance, worker safety and health, and employee attitude. There is a circular relationship, however, in which consequences are also causes. Careful attention to maintaining high product quality, machine performance, and human involvement is likely to be attended by excellent housekeeping.

OUTCOMES

What were the consequences of the policies employed in each of these cases? We do not feel that our data are sufficient to make unequivocal linkages between the methods of production system design and the results experienced on the factory floor. There are indications, however, that certain policies at the design and selection stage definitely contributed to the outcomes, and that there are lessons to be learned from the cases.

The results at Sikorsky illustrate the benefits of specific affirmative policies. As has been noted, the spindle/cuff machining-cell operators were involved early in the design and selection process—they came to be regarded as "invaluable consultants." An implicit policy prompted the involvement of the production employees. In the resulting system, 46 employees produced three times the output of the 60 people formerly required, and with one-sixth the number of reject parts. The system used no more space, halved throughput time and work-in-process, and had sufficient flexibility to more than double the output rate if pushed. There was little or no union resistance, but the mysterious stoppages that did occur indicated the need for even more, not less, communication with the operators.

By having to rely on outside machine suppliers for the major components in the cell, the Sikorsky engineers had to accept design flaws that made some of the work more awkward and less safe than was wanted, and expensive modifications were required after machines were installed.

Through teamwork among engineers, supervisors, and operators, Sikorsky's design team achieved a bootstrapped revolution in parts processing. Managers acquired a greater respect for the capabilities of the operators, and the operators gained pride in what they had been able to accomplish.

The Westinghouse case offers a useful contrast. The manufacturing cell was similar in nature to that at Sikorsky, but the approach taken by the system designers was significantly different. Although there was an implicit policy in the company to encourage operators to make suggestions during system design, there was a negligible amount of involvement in the FMS design. Operator participation was limited to quarterly design review meetings. Operators indicated that they felt they could have contributed to the design if given the opportunity.

Westinghouse's explicit policy was to restrict operator control to clearly defined parameters (the nature of the product forced similar rules in the Sikorsky spindle/cuff cell), and operator-initiated process modifications were restricted to establishing tool offsets when setting up a new part. Both companies had extensive skills development programs for operators; the difference between the firms appeared to be largely the degree to which the full capabilities of the operators were being actually recognized and used.

The outcome of the Westinghouse cell installation was considered a success, but it was significantly less remarkable than that at Sikorsky. After 18 months of operation the FMS was not completely integrated. Eight machines were operating independently in manual input mode, parts-handling logistics were awkward, and quality systems had not yet been fully incorporated. The system was not regarded by technical management as doing all that it should, and there was a sense of frustration on the shop floor.

These contrasting situations are mirrored in the evolution of assembly line design and selection policies at Whirlpool. Here we have four lines built in sequence over an eight-year period, with marked differences in policies employed and results achieved. The first of the lines (Line 3) was designed and installed with little or no employee involvement—the design objective was to minimize labor content through automation. Each operator had control over releasing product to the rest of the line and could stop the line if there were problems, but there was little opportunity for an operator to modify the process in any way or to perform maintenance on his or her workstation. There was no attempt to increase assemblers' skills—the goal was to eliminate the workers.

This first line had many problems during start-up. Parts variations that operators had been able to accommodate caused jams in the automated machines. The robots were slower than people. There was frequent downtime, and people had to be sent home. Widely separated workstations prevented the formation of social groups, made cross-training difficult, and blocked voluntary job rotation. Assemblers tended to be grim and did not like working on the line. Nevertheless, the line was technically successful, despite high downtime, scrap, and rework, but it took several years of modification to achieve a high-performance level.

The fourth line (Line 1), on the other hand, was started up with few problems, on time and within budget. Quality far exceeded expectations, and production targets were consistently met. The production team solved the daily crises, so management could focus on planning and strategy. When a reduction in schedule was imminent, the team of operators found other subassembly work to keep them busy.

What produced these differences? The case study team found that over the eight years Whirlpool's attitudes and policies had shifted from reliance on machines to reliance on people. The high-commitment team of engineers, managers, and operators that participated continuously in the design of the fourth line came up with a self-managing work group concept that was facilitated by the layout of the line. Each operator had control of line speed, could stop the line if necessary, and was encouraged to make modifications if there was team consensus. Operator capabilities were used in design decisions, problem solving, interviewing of job candidates, and making quality decisions. Design policies had evolved to the point where the production technology was designed (with operator input) to make use of all the abilities the operators brought to work each day.

The Polaroid and Timken cases have one feature in common: No operators were involved during the design or selection of either system. This, however, is where the similarity ends. The design objective for the engineers developing the viewfinder assembly line at Polaroid was to assemble a complex optical part in a hands-off, robotic assembly line. Human intervention was to be minimal. Timken designers, on the other hand, had the challenge of minimizing supervisory intervention in production operations by placing control of the process in the hands of the operators.

For the Polaroid designers the task was to integrate handling systems, assembly robots, and test devices under programmed control. This technical challenge took first priority. Because operators were not expected to intervene in the process, maintain it, or adjust it, the resulting machine design made intervention difficult, even dangerous. No tools were provided the operators, nor was any provision made for feedback or diagnostic information to the operators. Skill levels of operators were expected to be low, with low pay, high turnover, and little training.

The consequence at Polaroid was a system that was a technological triumph—when it ran. It was highly susceptible to jams that had to be cleared by operators, often at some risk. Operators actually took on maintenance and housekeeping tasks, including the use of tools, in order to keep the system running. Because the design essentially locked them out, operators had to devise ways of circumventing the controls so they could make adjustments or clear jams. Successful operation of the line was clearly dependent on the continual intervention of the operators, but this had not been recognized or provided for at the design stage. The resulting machine environment offered little opportunity for operator skill development or advancement, with a significant gap in job levels between operators and technicians in the area. Although it afforded a valuable learning experience for its designers, the viewfinder assembly machine was not the success it might have been.

Faced with the challenge of giving operators maximum control over the hot, dirty, risk-filled process of making steel, Timken designers proceeded down an entirely different design path from that at Polaroid. The degree of automation was not significantly less, but the Timken engineers designed the automated system with operator override, near-instant information feedback, and computer-aided decision support. Decisions were to be made at the operator level, so the system was designed to give the operator the information and means to do the job. Operators were afforded extensive skill development—each person was expected to be able to do every task required of his or her crew. First-level maintenance was an operator responsibility, and operators were encouraged to recommend process changes, with rewards from a gain-sharing plan.

These design policies, plus careful attention to the technical details of the process, gave Timken a highly productive, robust manufacturing system.The measures of success were many. Start-up was two months ahead of schedule and project completion $50 million under budget. Salable product was produced in the very first melt. Quality exceeded industry standards, and the process was consistent (3000 heats without a miss). Labor costs per ton were less than one-third those of industry norms. In addition to gaining highly favorable publicity on its performance, Timken had stolen a technological march on the rest of the low-cost alloy steel industry. The positive lessons learned from the Faircrest experience provided clear guidelines for technological changes introduced in other parts of the firm.

The Timken case demonstrates that operator involvement in design or selection of a process may not be necessary if company guidelines and policies are in place to shape the design. The principles guiding the Timken/Dravo design team kept the project on track. This experience does not detract from the lessons of the other cases. The good results at Sikorsky and Whirlpool, where employees were active participants, and the more negative consequences at Polaroid and Westinghouse, where there was no employee participation, indicate that production workers in a design or selection team can safeguard a project against inadequate or inappropriate design policies. In no case was employee participation a detriment. On the contrary, employee involvement and significant benefits to the company went hand in hand.

11

DESIGN PRINCIPLES FOR MACHINE–HUMAN COMPATIBILITY

We have identified the human factors policy areas that are pertinent in the design or selection of advanced manufacturing systems. We have examined how designers have treated human factors in a number of cases, and we have seen the outcomes. We are now prepared to distill from this study a series of design principles for engineers and manufacturing management that should be invoked whenever a new manufacturing system is being considered.

GENERAL PRINCIPLE

Most of these principles can be organized under the various policy areas we have identified. There is one basic principle, however, that applies across all policy areas:

Principle 1

Human beings will be operating, monitoring, and maintaining the production system being designed. These people will have normal human needs, feelings, and abilities, just as the designers have. The nature of their jobs, which may persist for years, will be an outcome of the design.

Designers who keep this simple truth in mind while a new project unfolds will take steps to ensure that the jobs they create in their new systems are deserving of human association. A well-designed production system will demonstrate respect for human dignity, ingenuity, and physical well-being.

Machine designers are designers of jobs. This simple fact is largely unrecognized in the training given to engineers, and it is seldom emphasized in the policies that guide engineers in their tasks. There is a clear need for greater emphasis on human factors education for technical people. Merely requiring engineering students to take a few "sociohumanistic" courses to round out their education will not do the trick. Carefully designed courses in industrial sociology or psychology are needed, taught by educators who are fully cognizant of the advancing technologies with which people will be working.

Because such courses generally do not exist, most companies will have to develop their own means of infusing this basic principle into their technical people. "Home-grown" training or indoctrination may prove to be as useful as any university-based course. Most companies will have had ample experience in designing or selecting new technology. A careful examination of the lessons learned in a few of these instances will prove enlightening, especially if operators and mechanics are encouraged to participate freely in that review.

Explicit policies stating the company's intentions regarding human factors in the design or selection of process technology can guide engineering action. Formal, written policies are preferable. We found many instances where there were unwritten, informal understandings that directed design teams. Such implied policies are unlikely to be followed consistently, however, and fewer resources will be allocated toward unrecognized or unstated goals.

Much has been written about human capabilities and needs in industrial settings. Seminal writers on this subject include Maslow (1943), Emery (1959), McGregor (1960), Trist et al. (1963), and Herzberg (1968). Abraham Maslow's hierarchy of human needs was instrumental in Douglas McGregor's development of his Theory X-Theory Y alternatives to human resource management. It also undergirded the motivational studies of Frederick Herzberg. Fred Emery and E. L. Trist (1960), of the Tavistock Institute of Human Relations in England pioneered in the concept of sociotechnical systems.

Richard Hackman and Edward Lawler (1971) defined three job characteristics that must be present if both individual needs and organizational goals are to be satisfied: (1) the worker feels personally responsible for a meaningful portion of the work, (2) job outcomes are intrinsically meaningful or experienced as worthwhile, and (3) there must be feedback about what is accomplished.

Each of our five case studies provides evidence of the correctness of these early viewpoints and of the correctness of this first principle. More recent additions to the insights into human needs and job needs can be found in articles by Blackler and Brown (1986), Walton and Susman (1987), Badham (1990), Ehn (1990), Kidd (1991), Brödner (1991), Wilson (1991), and Wobbe (1991).

COGNITIVE PRINCIPLES

The next ten principles relate to the seven cognitive human factors policy areas.

Principle 2

Whenever possible, involve production workers actively and early in the design and selection process.

Employee involvement is a concept that has gained wide currency in recent years and is probably the most important principle derived from our research. Generally, companies have involved workers primarily in organizational and production issues and not in equipment design. However, we find that experienced trained production workers can and will make significant contributions to design team deliberations. They can influence the effectiveness and efficiency of systems being designed, and they can contribute to the continued improvement of these systems.

Employee participation should not be regarded as a ploy to "sell" technology to the work force. Superficial employee involvement programs are not likely to succeed, as recent research indicates that such programs may not result in productivity improvement (Harrison, 1991). We found that when there was a deep commitment to include production workers in all aspects of the design and procurement process, there were dramatic improvements in the performance outcomes of the systems and the people.

This principle involves change in several directions. It requires breaking down organizational and attitudinal barriers first between engineering and manufacturing functions and, second, between professional salaried workers and hourly production workers. It requires involvement, exchange of ideas, and respect for all participants because of their potential contribution to system design rather than because of position.

In our study, the Sikorsky case is perhaps the most dramatic example of employee involvement in design and procurement. Workers were motivated to attend machine tool shows on their own time and made suggestions that resulted in major savings and significant improvements in quality. In addition to a general organizational climate that encouraged worker initiative, the key factor in this case was the early involvement of workers, long before the plan got off the drawing board. The fourth of Whirlpool's evolution of assembly line designs is another good example of early and continuous worker participation during system design.

Worker involvement in design should ideally begin in the initial stages of planning the production system.Although this may be difficult for a new process, there appears to be a significant payback. A dramatic case of employee involvement in all aspects of planning and production is reported in the Saturn factory in Spring Hill, Tennessee, which began production in 1991. There they formed a committee of plant managers, superintendents, union committee members, production workers, skilled trades workers and GM and UAW staff to design a new plant over a period of seven years. To procure production equipment they established a Product Development Team (PDT), which included manufacturing engineers, financial managers, material managers, quality engineers, and UAW technicians. The PDT visited potential suppliers and selected the firm with the best technology and best organization, and with whom they wanted to work in a supplier relationship. They found that this broad involvement not only resulted in much better procurement, but it also created a greater number of people who were experts in how the new technology worked. There was a broader knowledge base and much faster knowledge diffusion about a new process and technology, resulting in a steeper learning curve and better system performance. As one Saturn executive expressed it, their success "has largely come through letting our people drive the design process of our technologies" (Lewandowski and O'Toole, 1990, p. 31).

Worker involvement has been integral to a number of companies in European countries, perhaps most notably at Volvo. In fact, it became institutionalized in Sweden in a 1977 law that requires employee involvement in "decision making from the board level to the shop floor," and employee organizations are provided all financial information about the company (Gyllenhammar, 1977). Worker involvement, specifically in design, was demonstrated in two notable projects in the 1970s and 1980s, the DEMOS and UTOPIA projects in Scandinavia. In the UTOPIA project workers at all levels were involved in designing new computer-based newspaper page makeup and image-processing technologies. The projects were guided by a philosophy of democratic design. It resulted in technology designed to support a high-involvement work organization (Bodker et al., 1987; Ehn 1990).

The principles for worker involvement in design have been most significantly developed in software design, though largely neglected in hardware engineering. Although the technology might be considered substantially different, the underlying principles should be similar. In software, user involvement in design is considered paramount for the success of a system. This is a perspective that should be used in traditional hardware engineering as well. Another lesson from software design is the concept of viewing those working with the systems as "users" rather than just operators. They are thus considered as people using the technology as a tool to conduct their work rather than as mindless operators of a machine. Users know best what their job involves and their participation is thus considered vital to designing effective systems. Understanding end-user requirements is integral to the technical performance of the system, so user involvement is an important means of improving technology acceptance and use (see, for example, Rouse and Cody, 1987; and Levi, Siem, and Young, 1991).

It is important that workers be involved early in the process. For the design of a new process, this could require early selection of the work teams that are to be assigned to the new process, so one or more workers can spend time as a member of the design team. It also requires an attitude of acceptance and respect from engineering staffs who usually lead the design and procurement process. Promoting this principle of worker involvement can be accomplished through policy, education, and example.

Principle 3

Design the production system for operator control. Augment operator control with computer or programmed automation support.

The increasing interdependencies and cost of modern process technology make it important that machines be kept running a high proportion of the time. To keep automated manufacturing systems running, decision making must be done at the organizational level nearest the machines. In most cases, this means that the operator must be in control (Wilson, 1991).

This principle of operator control is most clearly demonstrated in the Timken Faircrest Steel Plant case. The design engineers were guided by a policy of putting decision making in the hands of operators, and they accomplished this by designing the needed tools into systems that facilitated such control. The plant has had a history of remarkably consistent output and quality.

When applied to automation technology, control seldom involves operators guiding tools or manipulating work pieces continuously. Instead, control implies intervention when conditions warrant. Control requires alertness, knowledge and judgment on the part of operators, and trust on the part of managers and engineers.

Companies sometimes make the error of assigning responsibility for a process to operators without giving them the means or know-how needed to control the process. Operators with real control over an operation are trained to understand the process, have easy access to the controls, and have near-instant feedback on the effects of their actions. They may also be assisted by look-up tables, computer simulation, and other computer aids (see Principle 5).

Davis (1971) describes the two functions left to human beings who work with automated or sophisticated technology: "deterministic tasks for which machines have not yet been devised, and control of stochastic events—variability and exceptions" (p. 439). The parts loader at the head of the Polaroid assembly line is an example of the former type of task. The person was performing a very routine job for which automation was too difficult and costly to design and build. The assembly operators on the fourth Whirlpool line had, as a group, control over the functions of the line, and over virtually all other aspects of their work routines. Obviously, they were performing many deterministic, routine tasks for which automation had not been designed, but they also had control over the nonroutine, unexpected, "supervisory" aspects of the operation.

With control comes commitment. Lund and Hansen (1986) mention a papermaking plant where the process had been designed to give operators "a major role in controlling the process. . . . Many of the plant operators were staying after work to play with the capabilities of the computerized process simulator. One of the problems for this firm was in getting its employees to go home" (p. 112). In this instance, operators had not only accepted the responsibility for seeing that the stochastic events were handled capably, but had taken on responsibility for improving the process through their ability to control the process.

The next two principles in the area of *feedback of production information* are closely related to the foregoing principle on control.

Principle 4

Timely information about the process must be available to the operator. This includes present operating conditions, results of past performance, and projections of future requirements.

If operators are to make decisions that control processes, they must have the information on which to act. As a minimum, operators should have access to local system performance data, but they should also obtain group or plant performance data that provides a broader context within which they work. Access to this more complete information demonstrates to operators that they are considered real members of a production team, not outsiders from whom information should be withheld.

Providing information to production workers is increasingly important as machines take over the actual execution of work tasks, and people become supervisors and analysts of the process. The Faircrest steel plant is a striking case of how production workers can use a broad range of information to improve the process. Providing business information such as material and operating costs to the operators helped them make production decisions that led to significant cost reductions. When the workers had access to information such as different material or electricity costs they were able to consider a broader range of factors that could improve production performance and product costs.

Traditionally, production workers have been provided with only limited information. The prevailing attitude has been one of allowing access to information on a narrowly defined "need to know" basis only. This restricts the range of decision making. The "need-to-know" criterion should be changed to "ought to know." Access to a broad range of information allows workers to expand their view of the production process and gives them another set of tools for conducting work in the "information age." Moreover, to motivate workers to feel real "ownership" of the process, management should provide them with information a responsible owner would expect to have.

Most computer-based systems have the information and the processing capability to present near-instant feedback of operating conditions and results. It remains only for the process designers to organize the information into readily usable form and to provide the means of presenting the information to the users.

Principle 5

In situations where operators are expected to have significant latitude of control or to exercise considerable judgment in making decisions, consider providing support in the form of expert systems, computer simulations, or similar aids.

The increasing complexity of equipment and processes, coupled with increased worker involvement and shop-floor decision making, may require quite sophisticated judgments. By providing decision-making tools, diagnostic aids, and easy-to-use interfaces, designers give workers the support they need. The Faircrest steel plant provides an example of a computer-based system that aids workers' production decisions. In the ladle-refining operation, a computer calculated a suggested alloy mix that the operator could use. Alternatively, the operator could have the computer calculate other mixes and then decide upon the one he or she felt was best. The system was designed to aid the operator, not supersede his or her judgment. It was thus viewed as a support tool rather than a deskilling automation system. It allowed workers to improve production in ways that went beyond the known algorithmic solutions for the process.

Other types of computer systems can be used to aid shop-floor decision making, control, and modification of the process. Programmable machine tools with graphical interfaces assist machinists and make adjustments much easier than traditional machines that often require machine code programming. For example, displays showing machine tool path simulations, in addition to digital displays, allow the machinist to view the changes before cutting metal and allows analog modification instead of exclusively digital input. In one machine tool design, data fields that require the worker to enter information are explained by graphical representation (Brödner, 1991). This approach helps production workers to translate their manufacturing knowledge into the formal specifications required for programming.

Principle 6

Design the system with "tools" that enable operators, technicians, or mechanics to modify the process easily.

Control over modification of a process once it is established (once a production run is set up) involves difficult decisions. It is almost impossible to design equipment that takes all contingencies into account (Rasmussen, 1986; Salzman, 1989). Modifications that are delegated solely to NC programmers or engineering staff, for example, may be significantly delayed until a specialist is available. Shortening the loop to allow for operator modification holds great potential for increasing machine uptime and output quality.

Opening up the system for modification of the process by operators cannot be done simply or without thought to system design and the support that nonengineering people may require. Modifications may require extensive knowledge about materials, product requirements, or customer specifications or procedures that are affected by the changes. In the spindle/cuff machining cell at Sikorsky, for example, some process changes required requalification of the product to meet government specifications.

Designing the system for modification by semiskilled or unskilled workers requires careful consideration of information needed, display design, and training. Mobilizing the skills of operators, mechanics, or technicians, while ensuring the quality and integrity of the product, involves at least three requisites: (1) in-depth training; (2) an efficient approval process for reviewing proposed changes, qualifying parts, and communicating change instructions to the organization; and (3) tools that make process changes easier and accurate.

Training must encompass not only *how* to run a machine but *why* certain processes are done as they are. Individuals qualified to modify a process should know how changes to the machine will affect the product, how the qualification process works, and what downstream effects these changes will have on inventory, product cost, and subsequent processes. Where a number of workers are involved in the same operation, or where there is a significant amount of turnover, it may be efficient to use videotaped instruction so everyone receives similar and complete instruction.

Procedures for controlling process changes may be mandated, as was the case in the spindle/cuff cell, but, even when requirements are less stringent, good shop discipline calls for orderly, systematic change procedures. When each individual follows a different pattern, it is almost impossible to track the source of process problems. Procedures should permit prompt emergency action, but should require documentation and follow-up to evaluate changes. Such documentation provides a database that increases learning by all. Operators can see what others are doing, and engineers can evaluate changes and spot problem areas.

The equipment designer's responsibility is to make certain that machines and equipment are "user friendly," so changes can be made readily. Process modification is most easily accomplished in advanced manufacturing systems by modifying computer programs. By providing teaching pendants, terminals, and keyboards with the systems on the manufacturing floor, designers give operating personnel the tools they need to improve the process.

At Timken the operator was given considerable latitude in controlling process variables, even though the basic steelmaking process did not lend itself to modification. Extensive computer-based decision aids allowed the operator to compare control decisions with computer-generated solutions. This built up confidence in the operators to the extent that they were able to suggest at least one basic process modification and make many suggestions for modification of computer displays and control algorithms. Prompt action was taken by engineers on these suggestions.

Principle 7

Design manufacturing systems to take full advantage of all the capabilities of people working with them.

This principle provides a real challenge to a technological design team. Not only will the design team wish to use the best that technology has to offer in the way of accuracy, speed, repeatability, endurance, adaptability, and strength, but it will also seek to use the best that people have to offer. Machine–human compatibility involves satisfying both the needs of technology and the needs of people.

The *needs of advanced technology* relative to people skills have been described by Lund and Hansen (1986) as:

1. Visualization—the ability to manipulate mental patterns, to apprehend what is happening within a process or series of processes, even from a remote control station.
2. Understanding of process phenomena—an appreciation for machine functions and machine–material interactions.
3. Conceptual thinking or abstract reasoning—the ability to solve problems as they arise, on the basis of understanding the process and interpreting information about the process.
4. Statistical inference—understanding trends, limits, and the meaning of data.
5. Verbal communication, oral and visual.
6. Attentiveness—alertness to nonroutine events or to changing system conditions.
7. Individual responsibility that safeguards the system and products, and that works toward continuous improvement of performance.

Conversely, on the basis of the research we have reported here and the research of others (Davis, 1971; Emery, 1959; Engelstad, 1971; Majchrzak, 1988; Taylor, 1971; Trist et al., 1963), it is possible to define a list of *human requirements* relative to advanced manufacturing technology:

1. Safety from injury, unhealthy conditions, stress, and physical discomfort
2. Opportunity to learn and grow in the job
3. Opportunity to understand and control the manufacturing system
4. Opportunity to solve problems and make decisions
5. Feedback on system and organizational performance
6. Relatedness to the product and to the organization
7. Recognition and reward

Most workers are able to bring to the workplace a willingness to exercise skills that answer, in varying degrees, the needs of advanced technology (the Lund-Hansen list). With training and performance feedback they can increase these skills and their proficiency in applying them. If design of the technology fails to provide appropriate worker interfaces (displays, for instance), the needs of the technology may not be satisfied. Machine performance is diminished. In addition, if worker initiatives are discouraged by locking the operator out of the process, machine performance is likely to be negatively affected. Emery (1959) found that workers prefer tasks that have a substantial degree of wholeness, where the individual has control over the materials and processes involved. Our list (which may yet be incomplete) incorporates factors that are embodied in this sense of wholeness.

The second list is not incompatible with the Lund-Hansen list. Advanced technology can be designed to meet the human requirements, just as humans can adapt to satisfy technological needs. Consequently, if process designers do their job right, there need be no mismatch between technological requirements and human needs and capabilities. The challenge in each instance of design or selection of new technology is to create the features that most appropriately satisfy all needs. Systems that meet this test will, as we have seen in the case studies, reward both the company and the individual. Thus, human requirements and technology requirements can be made not only compatible, but, through careful technological design, synergistic as well.

Principle 8

Design systems to encourage multiple skills development through training, job rotation, or informal job swapping.

As we have seen in the preceding discussion, there is ample reason to want workers to develop and use multiple skills in jobs associated with advanced manufacturing technology. New technology requires a greater range of skills. Human needs demand a wholeness and an opportunity to grow in jobs. New plant organizations are trending toward leaner, flatter structures and are pushing decision making to lower levels. Quality management programs emphasize the responsibility of production workers for product quality. All these factors, plus the trend toward worker teams, task sharing, job rotation, and joint problem solving call for broader skills and knowledge.

System design, then, must facilitate multiskilling. Machine designs should be accessible and understandable. This does not rule out complexity. It may mean, however, that the complexity becomes embedded in programmable logic controllers or computer programs that are essentially invisible to the user, in much the same way that electronic chips control automobile functions.

Accessibility implies operator knowledge of machine functions and means of feedback on machine conditions. When systems are integrated, information flows should also be integrated and accessible—workers on one machine, for example, can have access to information about the entire process or work cell. Workstations should be positioned sufficiently near each other so there can be interchange of information, job swapping, and mutual support. Central meeting places within the work area and use of the computer's communications capabilities can further facilitate development of a flexible work force.

In many production systems bottlenecks or flow stoppages occur at various points at unscheduled times. A robust system design will make it easy for workers to assist each other in resolving temporary bottlenecks by teaming up or reassigning work. Informal arrangements arrived at on the production floor will often provide a superior response (timing, appropriateness) to a crisis.

We found in our survey that the advent of advanced manufacturing technology in many companies brought with it a broadening of job descriptions and a reduction in the number of job classifications. The Sikorsky, Timken, Westinghouse, and Whirlpool cases all featured multiskilling as a policy. This trend is supported by other observers (Ginwala, 1987; Susman and Chase, 1986; Walton and Susman, 1987). Where multiskilling has been introduced, the result has been a more adaptable, flexible work force. This flexibility is useful not only in improving system performance, but it also permits changes in process technology to occur more readily.

Manufacturing system designs that isolate operators tend to discourage informal cooperation, problem solving, and on-the-job learning.

Principle 9

Avoid designing systems that "freeze in" or stratify jobs. If different levels of skills are required, design the system so there is a chance of progression from one job level to the next in reasonable steps.

Automation technology poses a threat to the moderately skilled worker. People in positions such as machine operator, welder, painter, inspector, and, increasingly, assembler can find that automation eliminates the skill aspects of the job. They can be relegated to loading and unloading work pieces, clearing jams, or stopping the machine when difficulties occur. The trend in high-skilled jobs, on the other hand, is toward further increases in skills requirements, as we saw in our survey.

If only low-skill and high-skill jobs remain, the danger is that the separation between the two levels will become too great for normal progression or advancement from the lower level. Lund and Hansen (1986) found this happening in instances of computer-based automation. It was also the situation for the crew on Polaroid's viewfinder assembly machine. We found a four- to six-step job-level gap between parts loaders and technicians, too great a distance for the loaders to leap.

The Whirlpool high-commitment line illustrates techniques for avoiding the skills gap problem. The line was designed to facilitate job switching, and minor maintenance activities were encouraged. Operators were given a broad-based training that covered problem solving, decision making, communications, and the relation of their jobs to the rest of the jobs in the plant. The resulting team of assemblers was enthusiastic, self-managing, and effective. Other companies have made similar discoveries through broad job classifications, multiskilling, and work teams (*Business Week,* 1986).

When only low-skilled jobs and high-skilled jobs exist in a system, people in the low ranks are discouraged from attempting to leap the gap that is created. Alienation and indifference can be the result. Broad job descriptions that encompass a wide range of skills, and payment for competence in those skills, are two approaches that break down barriers to progression.

Principle 10

Make it possible for operators to handle first-level maintenance and repairs on production systems. Provide essential tools, diagrams, supplies, and parts to accomplish the work.

As was stated earlier, expensive manufacturing systems must be kept running consistently and reliably. Machines repay their investment costs only when they are in operation and are making good product. The operator is usually first to notice a machine problem, and is immediately available to correct it if he or she is trained and has the means to do so. Keeping machines running should be part of operators' jobs, with support from mechanics, programmers, engineers, and other specialists.

Machine design should include consideration for the tools and other equipment needed to make routine adjustments, changeovers, and repairs. Operators should receive training in maintenance practices.

Computer monitoring and electronic displays provide instant information and rapid analysis of problems. Computer-aided diagnostics and trouble-shooting procedures greatly enhance the natural problem-solving abilities of operators. Computers can monitor system operation and notify operators when to perform scheduled maintenance, alert them to potential problems that are developing, and guide them in trouble-shooting and repair.

Computer-based guidance has been used, for example, for maintenance of automated letter-sorting machines. Interactive videodisc systems have been developed that allow the operator to call up information quickly and view video demonstrations of repairs or even instruction in basic operation. The system also allows operators to record "notes" about problems and fixes that have been encountered. These notes are then available for others to use when confronted with the same problem. Such systems support new skill development and decrease the number of occasions in which highly skilled or specialized technicians have to be called on to perform routine maintenance and repair. Improved machine utilization is a consequence.

Principle 11

During design and installation of a new system, create an environment that anticipates continuous improvement of the system by production personnel.

This principle has less to do with design of a system than with the behavior of the designers during the project.

Continuous improvement programs are prevalent in most leading companies today. Such programs tend to begin after new systems are installed and in operation. By establishing relationships of cooperation and trust with the production work force *during* design and/or selection of a system, designers can create the expectation among the production work force that continued suggestions and modifications will be received enthusiastically. When designers and manufacturing engineers become known to members of the work force, channels of communication are opened. Workers will feel free to introduce ideas that might never be expressed in a more highly structured organization.

In each of the case studies, a point was reached in the design project where schedule requirements forced designers to "freeze" their designs and get on with implementation. Recognizing that the designs were not perfect, the designers, in effect, anticipated further changes after installation. The small incremental changes that occur after start-up frequently make the difference between marginal and optimal machine performance. One major American engine company has stated that "improvement is part of the work" and expects employees throughout the organization to find ways to improve their own work.

Active involvement of operators, technicians, and mechanics in the design process sets the stage for later improvements. A foreman of the FMS at Westinghouse indicated that operators would not have had the knowledge to contribute much to the innovative design of the system, but he regretted that they did not have the opportunity to "live and grow with the system." In the case studies in which there was early involvement, designers gained an increased respect for the capabilities of the production personnel after working with them. Continuous contact between designers and operating personnel can also reassure the designers that their design intent is less likely to be violated in subsequent changes to a system.

The concept of continuous improvement demands a fresh perspective on the nature of production systems. No longer can a process be viewed as frozen and immutable. Instead, the process is treated as something that is always in transition toward ever-improving performance.

System design can be structured in ways that support this approach. The process can be made more open to change through modular design that permits change to one portion of the system without major alteration of the whole. One company designed its assembly line with an overhead conveyor that could be reconfigured quickly and easily. Only two pieces of equipment were bolted to the floor, and even those could be moved within 24 hours. All other equipment and line-side racks were on wheels so workstations could be changed or assembly steps relocated quickly.

Software programs, too, can be structured so as to facilitate changes. Making the technology more "transparent," that is, making it easily understandable to people at all levels of skill, invites suggestions for improvements. Finally, users can be given tools to help them develop new ideas: computer simulations, control data, terminals, programming devices, and, above all, training.

PHYSIOLOGICAL AND WORK ENVIRONMENT PRINCIPLES

In the *Physiological Needs* policy area the lessons from our cases are the source of several principles. These principles are intended to supplement the obvious principles that most companies follow regarding safety, health, comfort, stress, and ergonomics. These are somewhat less obvious principles that, nevertheless, are important in designing truly effective manufacturing systems.

Principle 12

Make the workplace so attractive to the worker that he or she will not seek means to escape it whenever possible.

"Progressive" companies take pride in the attractiveness of their lunch rooms, employee lounges, recreation areas, and classrooms. Companies that are even more progressive, however, will take pride in the quality of life they have made possible where people work.

If you design a manufacturing system or plant such that people leaving their work stations feel they are escaping from an oppressive environment, you sacrifice an opportunity to foster employee involvement and commitment. Much has been done to design appropriate office "landscapes" for staff workers. Similar attention needs to be paid to the factory.

It is rare to find a plant where this idea has been carried out. As part of the team assembly approach at Volvo's Kalmar plant, many of the assembly cells were arranged around the perimeter of the building in such a way that workers could see the outside through windows. IBM's Rochester, Minnesota, plant has maintained an attractive setting throughout—it has also been the winner of a Baldrige Award. There may be some connection.

Workplace design and workplace attractiveness are definitely areas where employee involvement can make unique contributions in this aspect of manufacturing technology.

Principle 13

Design for safety of human access and intervention for occasions when machines experience failure or stoppage. Use failure mode analysis and employee involvement to guide designs that enable safe correction of faults.

Machines fail. When they fail, people, often the operators, must try to get them back into operation. Guards that protect an operator during normal machine functioning can become hazards when a person is trying to remedy a problem quickly. This was demonstrated in the Polaroid case, where the company had gone to great lengths to establish procedures for safe equipment design. The pressure to meet production schedules is not an experience common to most equipment designers. They may have their own completion dates for delivery of machines, but the daily pressure to meet schedules is a way of life out in the factory.

Listen to operators' ideas of what can fail. Study guarding relative to the dynamics of the manufacturing process—the machine–material interactions, the machine–human interactions. How does one intervene to clear a jam quickly and safely? How does one realign a web of material while it is moving, without reaching into the machine? How does one clear waste or spills? Questions such as these require unique answers that demand as much ingenuity as any other part of the system design.

Provide operators with maintenance tools and training. If you do not, they will devise their own tools, or they will simply shut down the machine and wait until some mechanic can come to fix the problem. Neither alternative is very desirable.

If it is possible to operate a new system before introducing it to the production floor, it is often useful to ask an experienced operator or team of operators to make a trial run of several days to locate weaknesses in the design. Once a machine is in the factory, the pressure for output begins, and the opportunity for convergence of designers and operators on possible points of failure is lost.

Principle 14

Use ergonomic principles, but when in doubt make a mock-up and solicit advice from operators.

Ergonomics is the central principle of traditional human factors engineering. As we have used the term in this book, we have generally considered ergonomic principles as those that deal with the human system as a machine having certain capabilities and limitations, such as degrees of freedom of motion, tactile sensing, vision, and so forth. Most of these aspects of ergonomics will have a range of values that can be specified. In recent years, however, specialists in ergonomics have tended to broaden their field of view to include psychological or cognitive aspects as well. (Wilson 1991, for instance, talks about "cognitive ergonomics" and "social systems ergonomics.")

Three factors have made ergonomics considerations increasingly important to designers of manufacturing systems. The first is the high investment cost of advanced technology, which increases the potential loss that might occur from human error or ignorance. Second is the operating losses that can occur if there is excessive downtime. Human capabilities that can keep machines running properly and producing good product are essential to system performance.

The third factor is the increasing liability for medical and disability costs arising from injury because of faulty equipment design or application. In the past, back and shoulder injuries from overexertion and accidents were high on the list of disability costs in manufacturing. In recent years, however, we have become aware that even the modest motions in light assembly or in using a computer keyboard, where no great exertion is involved, can be harmful when performed repeatedly on poorly designed equipment. There have been dramatic increases in Cumulative Trauma Disorder (which includes carpal tunnel syndrome, first identified in 1717). It has been called the "occupational disease of the 1990s" by the Occupational Safety and Health Administration (Goldoftas, 1991).

Some large firms have made serious attempts to encourage engineers to incorporate ergonomic principles in their designs. IBM, for example, has prepared a small ergonomic handbook for use by its engineers. More comprehensive references are also available (Bailey 1982, Kantrowitz and Sorkin, 1983; McCormick, 1970). Software programs that model the human body and simulate human interaction with technology or work environments can assist the engineer during the design process. Three such programs are MANNEQUIN (Biomechanics Corporation of America), ADAMS/Android (Mechanical Dynamics, Inc.), and JACK (Center of Technology Transfer, University of Pennsylvania). MANNEQUIN is a PC-based software program that can display a variety of 3-D human figures of different genders, body sizes, and nationalities. These figures can see, walk, bend, reach, and grasp objects. Center of mass and torque computations can be made on the basis of how the figures interact with machines or other objects. The other two programs run on higher-powered computer workstations and provide a broader range of options for evaluating the static and dynamic interactions of people and machines. Aids of this type greatly increase the ease with which engineers can test their designs for ergonomic correctness and other performance criteria.

In certain instances where machine–human interactions are quite complex and unpredictable (as was true in the Timken case), the creation of a mock-up of a workstation is about the only way to fully evaluate the ergonomic requirements of the job. As we have found to be true with other aspects of new design, the involvement of experienced operators, mechanics, and technicians will provide insights that might not occur to an engineer.

Principle 15

Design for housekeeping.

"A place for everything and everything in its place." Cleanliness and order have an effect that extends well beyond the well-being of a worker. Employee attitude is influenced by housekeeping. Pride in workplace and pride in product go hand in hand (Hayes 1981). Machine performance and human performance are both dependent on the environment in which they work. A dirty, disorderly, workplace affects product quality, human safety, and machine reliability.

Both machine design and workplace design should avoid the accumulation of waste and should facilitate cleaning. Designers must provide for enclosed machine bases, automatic chip handling, and pneumatic removal of trim waste and dust. Design adequate clearances in machine sections where waste is likely to build up and must be cleaned out by operators. Provide operators with brooms, brushes, vacuum cleaners, and waste receptacles.

Provide storage cabinets or shelves for tools, supplies, and cleaning equipment. Use colors to make the workplace attractive, but use them also to highlight areas where dirt or waste collects.

Workplaces are often cluttered with unnecessary accumulations of work-in-process inventories. One of the great advantages of a "pull" or "kanban" system of material control is that it produces a radical reduction of goods-in-process on the factory floor (Schonberger, 1982). Highly flexible automated systems that have rapid changeover capabilities further cut inventories by accommodating smaller batch sizes. Group technology work cells reduce the amount of transport time and distances that parts must travel in the plant before assembly into finished product. All these approaches open up space and make the job of maintaining order much simpler.

As has been the case for many of our other principles, designing for ease of housekeeping is an area where production workers can provide expert advice. Designers who consult workers on this aspect of system design convey the message that housekeeping is important, and that each person in the company shares responsibility for it.

CONCLUSION

These 15 principles are not difficult for anyone to grasp. They can be easily taught. They ought to be included in the curriculum of every engineering student whose career is likely to include machine or product design. They should be part of the policies and procedures of all manufacturing companies, regardless of the degree of sophistication of their manufacturing systems. They should be studied by companies that supply machines or production systems to manufacturing companies.

Nor are the concepts new. All of them have been invoked in some way in human organizations over the years, but, as we have seen, they are rather rarely used during the equipment design and selection stages of manufacturing. This research suggests that it will be rewarding for engineers to use these ideas when they begin production system design.

It is one thing to understand a principle; putting it into practice is another matter. Ingrained attitudes and established policies are difficult to change. Many of these principles imply increased worker empowerment as a means of realizing greater worker contributions to the process. Our findings indicate that, when companies persist in following the old precepts of management prerogatives and control, they are destined to experience unnecessarily low limits on what they can gain from advanced technology and from their employees. On the other hand, companies that begin the process of employee empowerment where the jobs begin—in the design and selection of the technology—may well experience gains even beyond those they might have expected.

REFERENCES

BADHAM, RICHARD (1990). "Beyond One-Dimensional Automation: Human-Centered System Design in European Manufacture" in W. Karwonski and M. Rahimi (eds.), *Ergonomics of Hybrid Automated Systems-II.* Amsterdam: Elsevier.

BAILEY, ROBERT W. (1982). *Human Performance Engineering: A Guide for Systems Designers.* Englewood Cliffs, NJ: Prentice-Hall.

BLACKLER, FRANK, AND COLIN BROWN (1986, January). "Alternative Models to Guide the Design and Introduction of the New Information Technologies into Work Organizations." *Journal of Occupational Psychology.* pp. 287-313.

BODKER, SUSANNE, PELLE EHN, MORTEN KYNG, JOHN KAMMERSGAARD, AND YNGVE SUNDBLAD (1987). "A Utopian Experience: On Design of Powerful Computer-based Tools for Skilled Graphic Workers" in Gro Bjerknes, Pelle Ehn, and Morton Kyng (eds.), *Computers and Democracy.* Aldershot, England: Avebury.

BRÖDNER, PETER (1986). "Skill-Based Manufacturing vs. 'Unmanned Factory'—Which is Superior?" *International Journal of Industrial Ergonomics,* vol. 1, pp. 145-153.

BRÖDNER, PETER (1991, January). "Design of Work and Technology in Manufacturing." *International Journal of Human Factors in Manufacturing,* pp. 1–16.

Business Week (1986, September 29). "Management Discovers the Human Side of Automation." pp. 70–75.

DAVIS, LOUIS E. (1971). "Readying the Unready: Post-Industrial Jobs" in L.E. Davis and James C. Taylor (eds.), *Design of Jobs.* Baltimore: Penguin Books.

EHN, PELLE (1990, March). "Scandinavian Design: On Participation and Skill, Denmark" Paper presented at the Technology and Future of Work Conference, Stanford University.

EMERY, F.E. (1959). *Characteristics of Socio-Technical Systems.* Tavistock Institute of Human Relations document 527.

EMERY, F.E. AND E.L. TRIST (1960). "Socio-Technical Systems" in C. W. Churchman and M. Verhulst (eds.), *Management Science, Models and Techniques,* vol. 2. New York: Pergamon.

ENGELSTAD, PER H. (1971). "Socio-Technical Approach to Problems of Process Control" in L.E. Davis and James C. Taylor (eds.), *Design of Jobs.* Baltimore: Penguin Books.

GINWALA, KYMUS (1987). *Japanese Automation Worker Participation: Some Perspectives.* Cambridge, MA: M.I.T.–Japan Science and Technology Program, MITJSTP Report no. 87-12.

GOLDOFTAS, BARBARA (1991, January). "Hands that Hurt." *Technology Review.* pp. 42–50.

GYLLENHAMMAR, PEHR G. (1977, July-August). "How Volvo Adapts Work to People." *Harvard Business Review,* pp. 102–113.

HACKMAN, J.R. AND E.E. LAWLER (1971). "Employee Reactions to Job Characteristics." *Journal of Applied Psychology,* vol. 55, pp. 265–286.

HARRISON, BENNETT (1991, January). "The Failure of Worker Participation." *Technology Review,* p. 74.

HAYES, ROBERT H. (1981, July/August). "Why Japanese Factories Work." *Harvard Business Review,* pp. 56–66.

HERZBERG, FREDERICK (1968, January-February). "One More Time: How Do You Motivate Employees?" *Harvard Business Review,* pp. 53–62.

KANTROWITZ, BARRY H. AND ROBERT D. SORKIN (1983). *Human Factors: Understanding People-System Relationships.* New York: John Wiley & Sons.

KIDD, P. T. (1991, January). "Human and Computer-Integrated Manufacturing: A Manufacturing Strategy Based on Organization, People, and Technology." *International Journal of Human Factors in Manufacturing,* pp. 17–32.

LEVI, DANIEL J., CHARLES M. SIEM, AND ANDREW YOUNG (1991, July). "Using Employee Participation to Implement Advanced Manufacturing Technology." *International Journal of Human Factors in Manufacturing,* pp. 233–243.

LEWANDOWSKI, JIM AND JACK O'TOOLE (1990, March). "Forming the Future: The Marriage of People and Technology at Saturn." Paper presented at the Technology and Future of Work Conference, Stanford University, March 29.

LUND, ROBERT T. AND JOHN A. HANSEN (1986). *Keeping America at Work.* New York: John Wiley & Sons.

MAJCHRZAK, ANN (1988). *The Human Side of Factory Automation: Managerial and Human Resources Strategies for Making Automation Succeed.* San Francisco: Josey Bass Publishers.

MASLOW, ABRAHAM H. (1943). "A Theory of Human Motivation." *Psychological Review,* vol. 50, pp. 370–396.

MCCORMICK, ERNEST J. (1970). *Human Factors Engineering,* 3rd ed. New York: McGraw-Hill.

MCGREGOR, DOUGLAS M. (1960). *The Human Side of Enterprise.* New York: McGraw-Hill.

RASMUSSEN, JENS (1986). *Information Processing and Human-Machine Interaction: An Approach to Cognitive Engineering.* New York: North Holland.

ROUSE, WILLIAM B., WILLIAM R. CODY, AND KENNETH R. BOFF (1991 January). "The Human Factors of System Design: Understanding and Enhancing the Role of Human Factors Engineering." *International Journal of Human Factors in Manufacturing.* vol. 1.

ROUSE, WILLIAM B., AND WILLIAM J. CODY (1987). "On the Design of Man–Machine Systems: Principles, Practices, and Prospects" presented at International Federation of Automatic Control World Congress, Munich.

SALZMAN, HAROLD (1989, November). "Computer-Aided Design: Limitations in Automating Design and Drafting." *IEEE Transactions on Engineering Management,* vol. 36, p. 4.

SCHONBERGER, RICHARD J. (1982). *Japanese Manufacturing Techniques.* New York: The Free Press.

SUSMAN, GERALD I. AND RICHARD B. CHASE (1986). "A Sociotechnical Analysis of the Integrated Factory." *Journal of Applied Behavioral Science,* vol. 22 no. (3), pp. 257–270.

TAYLOR, JAMES C. (1971). "Some Effects of Technology in Organizational Change." *Human Relations,* vol. 24, pp. 105–123.

TRIST, E. L., G. W. HIGGIN, H. MURRAY, AND A. B. POLLACK (1963). *Organizational Choice.* London: Tavistock Publications.

WALTON, RICHARD E. AND GERALD I. SUSMAN (1987 March/April). "People Policies for the New Machines." *Harvard Business Review,* pp. 98–106.

WILSON, JOHN R. (1991, July). "Critical Human Factors Contributions in Modern Manufacturing." *International Journal of Human Factors in Manufacturing,* pp. 281–297.

WOBBE, WERNER (1991, July). "Anthropocentric Production Systems: A Strategic Issue for Europe." Anthropocentric Production Systems research paper series. Brussels: Commission of the European Communities, vol. 1.

INDEX